Monstrous Politics

Critical Mexican Studies
Series editor: Ignacio M. Sánchez Prado

Critical Mexican Studies is the first English-language, humanities-based, theoretically focused academic series devoted to the study of Mexico. The series is a space for innovative works in the humanities that focus on theoretical analysis, transdisciplinary interventions, and original conceptual framing.

Other titles in the series:

The Restless Dead: Necrowriting and Disappropriation, by Cristina Rivera Garza

History and Modern Media: A Personal Journey, by John Mraz

Toxic Loves, Impossible Futures: Feminist Living as Resistance, by Irmgard Emmelhainz

Drug Cartels Do Not Exist: Narcotrafficking in US and Mexican Culture, by Oswaldo Zavala

Unlawful Violence: Mexican Law and Cultural Production, by Rebecca Janzen

The Mexican Transpacific: Nikkei Writing, Visual Arts, and Performance, by Ignacio López-Calvo

CRITICAL
MEXICAN STUDIES

Monstrous Politics

Geography, Rights, and the
Urban Revolution in Mexico City

Ben A. Gerlofs

Vanderbilt University Press
Nashville, Tennessee

Library of Congress Cataloging-in-Publication Data
Names: Gerlofs, Ben A., 1985– author.
Title: Monstrous politics : geography, rights, and the urban revolution in
    Mexico City / Ben A. Gerlofs.
Description: Nashville, Tennessee : Vanderbilt University Press, [2023] |
    Series: Critical Mexican studies ; volume 7 | Includes bibliographical
    references and index.
Identifiers: LCCN 2022012492 (print) | LCCN 2022012493 (ebook) | ISBN
    9780826504777 (Paperback) | ISBN 9780826504784 (Hardcover) | ISBN
    9780826504791 (epub) | ISBN 9780826504807 (pdf)
Subjects: LCSH: Historical geography. | Ethnology—Mexico City. |
    Urbanization—Mexico City. | Politics, Practical—Mexico City.
Classification: LCC G141 .G48 2023 (print) | LCC G141 (ebook) | DDC
    911/.7253—dc23/eng20221025
LC record available at https://lccn.loc.gov/2022012492
LC ebook record available at https://lccn.loc.gov/2022012493

For my mother, Jacqueline Kay Gerlofs (née Lenderink), masterful and unsparing practitioner of transgressive humor and unashamed lover of Casa de Toño.

Contents

Figures

Acknowledgments

In conducting the research for this book and in the writing, I am thankful to have had incredible intellectual, emotional, logistical, financial, caloric, and other support from many corners within and beyond the academy. For these kind souls and generous institutions, whose geography has sprawled across much of the globe since the process began, I will be forever grateful.

At Rutgers University, Kathe Newman pushed me to think more carefully about the operation of power in urban geography, and to produce writing that lives up to my auspicious titles. Don Mitchell has been patient and supportive through intellectual and personal challenges, and I am forever realizing how influential his work and his approach have been on my own. I am always in conversation with Don, even when he doesn't know it, fulfilling his maxim that former students never are. Bob Lake more than lived up to his reputation as a wonderful advisor, and his patience and support are much appreciated, as are his timely words of wisdom. Bob continues to gently help me figure out what it is I'm really trying to say, and help me say it better. Asher Ghertner has likewise been the very best of advisors. His advice kept me on the yellow brick road during some rocky times, and helped me find it again when I wandered off. His incredible ability to cover a seemingly infinite field of knowledge remains impressive and inspiring, and I find his influence in many of my most-prized pedagogical practices as well. Rick Schroeder also deserves immense thanks for being patient with my dogged attempts to make sense of dialectics. I was also privileged to work with an incredibly talented group of graduate students and friends at Rutgers, especially Sangeeta Banerji, Hudson McFann, David Ferring, Mónica Hernández Ospina, Priti Narayan, Sean Tanner, Rich Nisa,

Evan Casper-Futterman, Brian Baldor, Juan Rivero, Ben Teresa, Ana Mahecha-Groot, Marlaina Martin, Thomas Crowley, Wei-Chieh Hung, David Eisenhauer, Helen Olsen, Erin Royals, Jonah Walters, Ali Horton, Debby Scott, Ally Sobey, Stuti Govil, Dawn Wells, Sadaf Javed, María García, Mike Brady, Diya Paul, Ariel Otruba, Jenny Isaacs, Natalie Teale, Jack Norton, Josh Randall, and Asher Siebert. I am particularly grateful to those who gave feedback on versions of several chapters as part of Asher Ghertner's urban lab (including brilliant visitors Tom Cowan and Harry Pettit). I also benefitted from the guidance and comradery of several other faculty members, including especially Laura Schneider, Daniel Goldstein, Nina Siulc, Trevor Birkenholtz, Kevin St. Martin, and Kevon Rhiney. I owe a special thanks to Mazen Labban, who remains an incredibly generous and brilliantly insightful interlocutor and friend.

I am grateful to colleagues and friends at Dartmouth College for their tremendous support during my year teaching in the Department of Geography, especially Mona Domosh, Frank Magilligan, Tish Lopez, Kate Hall, Richard Wright, Chris Sneddon, Josh Cousins, Jason MacLeod, Susanne Freidberg, Ryan McKeon, Abigail Neely, Coleen Fox, Laura Ogden, Greta Marchesi, Elizabeth Wilson, Garnet Kindervater, Stephanie Spera, and Garrett Dash Nelson.

Princeton University's Program in Latin American Studies (PLAS) was a wonderful place from which to think and rethink; conduct fieldwork; share and received feedback; and to research, write, and/or edit much of this book, and I am extremely thankful to PLAS and to Gabriella Nouzielles for inviting me to join them as a postdoctoral research associate. PLAS is a vibrant, stimulating, and very welcoming community, made all the more so by their fantastic staff: Rebecca Aguas, Damaris Zayas, Jeremia LaMontagne, Eneida Toner, and Director Gabriela Nouzeilles. Thanks to Ryan Edwards for insights, camaraderie, and sharing an office (even though I often work with the overhead lights off, like some kind of vampire), and to Jessica Mack, Aiala Levy, Noa Corcoran-Tadd, Bridgette Werner, Miqueias Mugge, Leonardo Cardoso, Monica Amor, Farraz Felippe, Eduardo Moncada, Rafael Sánchez, Ana María Ibañez, João Biehl, Cristina Freire, Jeremy Adelman, Marcelo Medeiros, Arturo Arias, Mario Gandelsonas, Cecilia Fajardo-Hill, Rachel Price, Fernando Acosta-Rodríguez, Vera Candiani, Aaron Shkuda, and Arturo Alvarado for sharing Princeton and PLAS with me, and for so many great events and stimulating conversations. I am grateful to Doug Massey and Gabriela Nouzeilles for mentorship and support, and to Javier Guerrero and especially Rubén Gallo for providing helpful feedback on an early chapter of this book.

My new colleagues within and beyond the Department of Geography at the University of Hong Kong have been extremely welcoming and supportive (especially with a harried arrival amid the global COVID-19 pandemic), and have made Hong Kong a fantastic place from which to finish the manuscript's final edits. In particular, I am thankful to George C. S. Lin, Becky P. Y. Loo, He Wang, Ben Iaquinto, Patrick Adler, Frank van der Wouden, Mia Bennett, Junxi Qian, Nicky Y. F. Lam, P. C. Lai, Jim Lenzer, Peter Koh, Yongsung Lee, Wendy Chen, Steven H. S. Zhang, Calvin Tribby, Lishan Ran, Zhenci Xue, Jinbao Li, Yanjia Cao, Jimmy Li, Wes Attewell, and Peng Gong. Outside of the department, I am thankful to Monica Lee Steinberg for both comedic theory and comedic relief, and most especially to Tom Barry for invaluable comradery and insightful morning hikes.

Roderic Ai Camp generously provided critical feedback on one of the book's chapters (Chapter 5), as did Mustafa Dikeç (on an earlier version of Chapter 4, along with three anonymous reviewers) and Ronan Paddison (on an earlier version of Chapter 3, along with three anonymous reviewers). Others who have shared specific knowledge, offered guidance or assistance on particular aspects of the research, or chatted about chilangolandia over a friendly meal or beverage include (to name only a few): Lorena Zárate, Silvia Emanuelli, Nick Crane, Priscilla Connelly, Veronica Crossa, Luis Alvarez León, Sergio González, Arturo Cadena, Pablo Gaytán Santiago, Jeronimo Díaz Marielle, Jaime Rello Gómez, and Enrique Ortiz Flores. I am grateful for the help and support of my editor, Zack Gresham, and series editor Ignacio Sánchez Prado. The book also benefitted immensely from the helpful feedback of two anonymous reviewers. A large portion of Chapter 3 and a small portion of Chapter 5 were previously published by *Urban Studies*, and much of Chapter 4 was previously published by the *International Journal of Urban and Regional Research* (IJURR). I am grateful to Sage and Wiley, respectively, for permission to republish these materials.

For research and travel support I am very thankful to the Department of Geography at Rutgers, the Centers for Global Advancement and International Affairs (GAIA) at Rutgers, the Center for Latin American Studies (CLAS) at Rutgers, the Graduate School of New Brunswick at Rutgers, the MaGrann Foundation, the Urban Geography Specialty Group of the American Association of Geographers, Habitat International Coalition-Latin America (HIC-AL), the Department of Social Geography and the broader Instituto de Geografía at the Universidad Nacional Autónoma de México (UNAM), and the Program in Latin American Studies (PLAS) at Princeton University. In particular, I wish to thank the many brilliant and wonderful people at HIC-AL (especially those already named above) for all of their introductions, advice, guidance, and friendship; Naxheli Ruiz Rivera at UNAM for her patience and helpful insights; Virginia Jabardo for sharing her office (and her friendship); and Amy Lerner for too many things to list, from Brooklyn to Mexico City. I am also grateful to Neil Brenner and Carlo Diaz for facilitating a short research visit at the newly established Urban Theory Lab at the University of Chicago, where I made some of the final edits in the summer of 2022.

I also wish to thank my interviewees and other research participants, most of whom will remain anonymous, and the countless people who shared even a few words with me in various places across the city, all of which were more useful than those folks can possibly know. These thanks apply doubly to those many persons forever hustling to make ends meet who chose to spend a little of their precious time helping me understand the city's rhythms and the lifeways of its fascinating inhabitants.

My friends in Mexico City, only some of whom would likely recognize themselves in this work, are also owed great thanks. Among them are Mario, Ian, Liam, Carolina, Molly, Carlos, Lana, Herón, Jesús, Lily, Lalo, Ana, Patricia, Lara, Gael, Joaquín, Ricardo, Yuko, and Marí. A very special thanks to Alex and Claire, both of whom opened a great many doors and helped in innumerable other ways, and whose friendship is highly prized.

To my family, Jackie, Mark, Annie, Isaac, Honor, Frankie, Torie, Luke, Meg, Dan, Evan, Michelle, Lindsay, Chris, Eloise, Christopher, Danny, Judah, and all my innumerable cousins, aunts, uncles, grandparents, and assorted others: I love and appreciate you all, and will be forever grateful for your love and support throughout this process. Yes, I'm still studying geography. And no, that's not "rocks," exactly.

Ignacio, my "Mexican wife"—by your own description—and most treasured friend and companion, I owe you more than you can know, and I only hope that a lifetime of amity and occasional rescue from the wilds of your urban dionysia will be enough to show you what your friendship means to me. Though I didn't know it when I started this project, none of it would have been possible without you.

Marissa, my partner and wife, your love and support have enabled me to ride out the rough patches and stay on course through trials, and your encouragement has meant the difference between success and failure more times than I could count. You've always been there to push me or to help me along, and to knock over the little toy animals when I need a reality check. I love you and I thank you.

Introduction

This is not the moment to analyze our profound sense of solitude, which alternatively affirms and denies itself in melancholy and rejoicing, silence and sheer noise, gratuitous crimes and religious fervor. Man is alone everywhere. But the solitude of the Mexican under the great stone night of the high plateau that is still inhabited by insatiable gods, is very different from that of the North American, who wanders in an abstract world of machines, fellow citizens and moral precepts. **Octavio Paz (1985, 19–20)**

Tepito es como un orgasmo, porque todo el mundo habla de él, pero poca gente lo conoce. **Lourdes Ruiz Baltazar**

As is often the case, the study that became this book began with a profound paradox. The operative premise was that for a brief period around 2010, the notion of *el derecho a la ciudad* (the right to the city) had held significant political purchase in Mexico City. In those days, most city officials knew and many heralded the phrase, prominent nongovernmental organizations (NGOs) and other civil society groups used and lauded the idea, and respected social movement leaders and an impressive breadth of other factions of the city's grassroots promoted it with a zeal unmatched by most other banners. The city's *jefe de gobierno*, Marcelo Ebrard, commissioned a charter based on this idea and tailored to the Mexican capital, which he and other officials signed in a highly anticipated public ceremony.[1] But by 2015, when I arrived in the city to conduct ethnographic fieldwork with the movement pursuing and promoting the right to the city, the phrase had all but vanished from public discourse. When the next *jefe* was dealt a severe and unanticipated political blow by residents in that year, for instance, the right to the city was conspicuously absent from public events and private meetings, traditional and social media, and promotional materials and related ephemera. Despite having achieved an impressive corpus of municipal instantiations and public endorsements, the right to the city had not achieved the stepping stone, benchmark, or platform status that academic and popular proponents and the concept's own recent political trajectory might have suggested. The right to the city's want of appearance in what became a momentously significant political process was deeply perplexing. As would become clear, this apparent demise was inescapably entangled within a broader set of processes and geographies in ways that raised a host of questions about the nature of grassroots politics, social movements, and *plataformas*, and the shifting landscapes of state-civil society relations in contemporary Mexico City.[2] The city's last hundred years serve as a powerful, even controlling filter for how the city's residents, political operators, and others perceive the political present, often in ways that are neither fully elaborated nor fully appreciated.[3] Understanding how contemporary militancies make sense of and act within and upon the city's complex political cartographies, in other words, requires a deeper engagement with the historical geography of Mexico City and the socio-spatial and political-economic dimensions of its rapid urbanization over the course of the last century. This book is the result of that inquiry.

The transformation of the Mexican capital over the last hundred or so years amounts to nothing less than an urban revolution, whether considered from the vantage of the city's achievements of political primacy, domination, and emancipation; its demographic, architectural, and environmental embodiment of the daydreams and nightmares of urban age and peak urbanism theorists; or the arc and through lines of the most compelling everyday narratives of its political and geographical development.[4] Mexico's two most mythic creations of the period are undoubtedly the Partido Revolucionario Institucional (Institutional Revolutionary Party, hereafter PRI) and its capital city. Both have long been painted with the brushes of monstrosity, and both have borne the brands (and often convincingly played the roles) of leviathan across countless genres.[5] The throes of their conflictual and co-constitutive development since the Mexican Revolution begun in 1910 violently transfigured the city and the urban consciousness of its populace, slicing and stacking new layers into the palimpsest. Sensations of this history particulate and palatial are everywhere apparent in the contemporary city, from the glorified busts and imperious architecture of its grandest palaces, plazas, and promenades, to the insurgent informalities that stubbornly populate the edges of even these same spaces, and from the booming sirens and shrieking wheels of the astonishing feats of infrastructural engineering that keep the city astride the beckoning chasm of environmental collapse to the dust of the city's atmosphere that continually, persistently coats each and every surface of the Aztec Metropolis with a subtly occluding gray film, from the clothes hung out to dry to the respiratory pathways of its more than twenty-million inhabitants. It is from within this milieu that contemporary residents operate, weighted down and lifted up by their century of (institutional) revolution. Through this twisting and winding century, I argue, a revolutionary structure of feeling has developed among the citizens of chilangolandia.[6] This structure of feeling describes a crucial mode of refraction for contemporary residents, the means by which both the weight of the past and hope for the present and future are organized and understood. In other words, it is a crucial heuristic for understanding the structural, institutional, spatial, and political transformations of the last century in Mexico City and the ways that contemporary political actors make sense of and work within these socio-spatial conditions.

DIALECTICAL INVESTIGATION IN THE CITY OF *ALBURES*

I approach and present this material with deference to the plurality that is and has always been Mexico City. The metropolis now called Mexico City has been many things to many peoples over the course of the last millennium. To the Aztecs, it was a defensible position and eventually the seat of an impressive empire. To Cortés and the *conquistadores*, it was the crown jewel of what would become New Spain, already a thriving metropolis requiring only the gilding of a Spanish baptism. Generations of Mexican and European rulers in Cortés's wake found a similar purchase in the city, a means by which a backward or stunted country might find the saving grace of civilization or modernity. To foreign visitors in the wake of European conquest, the city has routinely been said to evoke unexplored depths of emotion. The wonders of its natural splendor and civic and cultural achievement are matched only by the horrors and ravages of its twentieth-century development, however they might be experienced. It has been conquered and reconquered many times over since it was established as the Aztec capital, and continues to be an object of desire for politicians and their parties, much as it once was for conquerors and *caudillos*. It is the beating heart of a vibrant and diverse country whose culture and history refuse easy qualifications, whose biomes and landscapes contain some of the planet's rarest plant and animal life, and whose climatic range boasts an impressive spread of Köppen-Geiger classifications, from its dry alpine peaks and immense northern deserts to its dense Yucatán jungles and the many natural harbors spread across its vast Caribbean and Pacific coastlines. Sitting near the center of the country some four degrees below the Tropic of

Cancer, Mexico City's environment proves true to national form as the site of a humid subtropical highland climate with famously wet summers and flora and fauna that are truly splendid, like the curious *Axolotl* found only in the remnants of the great lake in Xochimilco in the south of the city, whose adorable permanent smile and all-around infectious cuteness caused a viral spike in interest in the critically endangered species in recent years. To generations of migrants from Asia, Europe, Central America, and Mexico's own hinterlands, the city has signified the potential for economic and social opportunity, and it continues to entice such pilgrims with precisely this promise. Now the largest city in the Western Hemisphere, the Aztec Metropolis is home to some of world's wealthiest and poorest people, as are so many of the rapidly expanding cities of the Global South. Its environmental problems are quite literally the stuff of legend, and its reputation for political corruption and violence are well earned. Its artistic and culinary traditions, among other draws, have moderated such national and international appraisals, especially in recent years, and the city is now widely considered a global destination for both leisure travel and investment.

Few theoretical or methodological languages are capable of productively engaging simultaneously with the nascencies of Mexico City's social, political, economic, and environmental ills, the ever-expanding sets of solutions and agendas proffered to address them, and conflictual fault lines that open up around and through these issues as interests compete for all manner of influence in the course of the city's caustic dance on the edge of catastrophe. As I am interested in exploring both the historical and geographical roots and contemporary manifestations and contestations associated with Mexico City's urban revolution, I have adopted an integrative and recursive methodological posture, which is reflected in the structure of the book. Part I is organized as an historical geography of the capital from the last years of Porfirio Díaz's regime on the cusp of the Mexican Revolution to the present, and Part II is anchored by ethnographic explorations, each addressed to a trajectory of conflict derived from this historical geography. Blending archival and ethnographic materials allows historical legacies and contemporary fault lines to inform one another, coalescing into a unique perspective on the tumultuous past and uncertain future of Mexico City and its urban revolution. To accomplish this integration, I rely on a diverse set of qualitative methods and evidentiary materials collected over several years, including semi-structured interviews, media reporting, archival materials (especially as provided by NGOs and other civil society organizations, and by the archives of several leading newspapers), legal and political documents and published opinions, public releases, secondary social-scientific and humanities literature, demographic and electoral data, photographs, expository cartography, and ethnographic fieldnotes of extensive participant and passive observation.

This methodological posture is grounded in my understanding and practice of dialectal materialism, following most particularly the mode of abstraction laid out by Ollman (1971, 1993, 2003).[7] This approach has had certain consequences for both the mode of investigation and of presentation, though perhaps not always in ways that will be immediately apparent to the reader. As Marx (1967b) once explained of his own method:

> Of course the method of presentation must differ in form from that of inquiry. The latter has to appropriate the material in detail, to analyse its different forms of development, to trace out their interconnection. Only after this work is done, can the actual movement be adequately described. If this is done successfully, if the life of the subject-matter is ideally reflected as in a mirror, then it may appear as if we had before us a mere a priori construction.

Thus, it is, for example, that the revolutionary structure of feeling I introduce as an analytical apparatus in Chapter 5 may appear as an invention spawned from misguided conceptual "coquetry," as Marx might have it, of the grandest idealist design. What goes unseen in this mode of presentation, however, is the wide-ranging and incessantly iterative analytical process by which this guiding principle was discerned. It would of course have been possible to follow any number of other analytical avenues in seeking to understand the process of political reform in Mexico City, but each would have brought a different set of connections into focus and illuminated or obscured various explanatory or contextual factors. In this case, it is my intention to provide a basis for understanding the movement of political reform and the particular events, documents, and arguments pertaining to it from not only a political but also a social perspective. That is, as I argue that understanding political reform requires abstractions spun from a somewhat lengthier temporal extension than are typically on offer— so as to include its inescapably significant social dimensions—such factors as public reception ought not simply to drop off at the end of the story but rather should form part of its frayed and incomplete ending, however difficult it may become to provide a coherent composition to the overall narrative. Arbitration of my successes and failures in this torturous endeavor, I leave to the reader.

Putting aside this disclaimer on the mode of presentation, the most important aspect of this dialectical methodology to be laid out at the outset is the function of abstraction. Abstraction, according to Ollman (2003, 60), deals with "the intellectual activity of breaking [a] whole down into the mental units with which we think about it," a process that necessarily does a certain violence to the world as it actually exists. Following the method Ollman carefully elaborates, I take as axiomatic the notion that no thing, person, party, relationship, event, etc., is ever only what it appears, however faithful a surface appearance may be to what it purports to represent. That is, this is not simply a dichotomy of a necessarily partial appearance and an underlying reality. Rather, the same "thing" is instead conceived as a relation, or rather a series of relations. Each such relation can take on any number of appearances when viewed from different angles, with each perspectival adjustment bringing different connections into and out of focus. Changing the angle of approach in this sense does not change some ontological essence that exists apart from the position of the analyst. Rather, this analytical position rejects such a premise of ontological discreteness. "Relations," as Ollman labels such relational totalities (collections of relations particularly named), appear differently based on the angle of approach, among other factors, in part because the way a relation is approached is one of the many factors of its functional existence (an approach to dialectics to be more fully elaborated in Chapter 3). One way to demonstrate the import of this approach is to illustrate the practical utility of casting abstractions that are sufficiently broad as to capture their relevance to a given issue, relationship, or investigation. For example, Mexico's Partido de la Revolución Democrática (Party of the Democratic Revolution, hereafter PRD) was founded in 1989, and many academic and popular treatments of the party begin their analysis of this political party at this time or shortly before, with its origins in the Frente Democrático Nacional (National Democratic Front, hereafter FDN). Such a treatment will bring into focus the relations between this party and the circumstances surrounding its inception, including its participation the 1988 presidential election (as the FDN), its relationship to the parties that formed the FDN, and the party to whom it ultimately lost the election (the PRI). Other treatments will cast a broader abstraction, setting an inquiry temporally further back in the 1980s and bringing into focus the tensions between what would eventually become the FDN and the PRI which still housed many of these elements, the PRI's internal factional tensions, and the macroeconomic policies and events that greatly exacerbated these trends. Still other treatments might reach even further in time, to include the stable decades of PRI leadership during which many of the leading members of what would become the PRD were trained and during which some of these persons established enduring (patronage) networks throughout Mexico City and the country more generally. None of these three approaches are precisely correct, in the sense of providing a faithful representation of the essential PRD that is somewhere awaiting its ultimate discovery. Rather, each abstraction brings different connections into focus, with different analytical possibilities. The first choice may be most useful for probing the dynamics of political pluralism and electoral reform in the 1990s, leading to the first presidential opposition victory since the Mexican Revolution in the 2000 elections. The second may be most useful for exploring the tensions between the PRD and the PRI, or between the PRD and its own breakaway party, Morena, which would go on to win the 2018 executive elections both locally and nationally. The third may be most useful for understanding the particular character of PRD practices and party operation in the city, especially with respect to electoral manipulation, graft, clientelism, and other tactics and strategies traditionally associated with the PRI. With each of these temporal abstractions of the PRD, different connections and relationships come into view, each expressing potentially compelling explanations for the behavior of PRD operatives in the city's poorest neighborhoods, for instance, or for the rumors of an alliance between PRD and PRI executives in a presidential election. To mislay even one aspect of this process of abstraction, temporal extension in this example, would obscure such relationships.

Quite aside from its theoretical and methodological advantages, and arguably of far more significance, this practice of dialectics most closely approximates the way that language and meaning are bent and manipulated in the everyday rhythms of chilangolandia.[8] Few words spoken in Mexico City have only one referent, especially when the wide world of kinesics (body language) allows for practically infinite variations of inflection. To be sure, this is a general principle of human communication systems. Mexico City-Spanish, chilango Spanish, however, is (in)famous for its interplay of simultaneous meanings.[9] Language games of masculinity, sexual dominance, political sarcasm, and many more themes are openly discussed and implicitly or explicitly played, sometimes ending in raucous laughter and sometimes in physical violence, especially among men. After all, Mexico City is the land of the *albur*, a hypersexual joke, insult, or social challenge built from a pun, a phonetic manipulation of a name or common saying, or some other twisting or bending of a phrase to give it a sexual, emasculating, or otherwise provocative meaning. This is a language that (especially male) foreigners, including Mexicans from outside the city, are encouraged to learn quickly if they are to avoid public ridicule. The capital's residents have also invented so many quotidian uses for the verb *chingar* (to fuck) that a comical (and comically large) usage dictionary (the *Chingonario*) is sold in tourist shops and bookstores throughout Mexico City, and now even abroad. Practically, shades of meaning simultaneously spoken in different registers routinely run through conversations, and language is commonly understood as a highly expressive and flexible form of communication. Words, phrases, inflections, pitches, rhythms, etc., bend and twist the meanings they carry, producing an emotive current beneath the superficial flow of words to inflict pain, convey love, make a filthy joke, or rearrange power dynamics, often in combination. This is not, of course, to say that nothing can be accurately understood amidst this shifting maelstrom. Rather, as even a *gringo* may easily learn, the field of meaning in Mexico City is best approached not as an archeological dig, where inert treasures are carefully discovered and delicately handled, but rather as a dance in which partners are constantly exchanging and the keys are to relate to your partner and to move within the flow of the music. This approach to language complicates any simple picture of Mexico City, and has too often led to appraisals of a duplicitous intent to mask a singularly "true" reality with flowery language or political pageantry. Though there is doubtless plenty of such behavior to be found (including in the chapters that follow), I argue that Mexico City is best and most fully understood by means of an approach attuned to deciphering the infinite shades of meaning that may exist between an absolute truth and a bald-faced lie.

OUTLINE OF THE BOOK

Part I comprises Chapters 1 and 2, which present an historical geography of Mexico City from just before the Mexican Revolution of 1910 through the present. Both the rationale for and construction of this historical geography developed organically through ethnographic study, as the social gravity of many of the events and processes elaborated in Part I was initially impressed upon me by friends, neighbors, colleagues, and research participants of all kinds in the course of my research on contemporary mobilizations. In these two chapters, I have systematically reconstructed this historical geography—which is so often conveyed in fragmentary and messy ways, burdened by the persistence and impermanence of memory and the incalculable influence of narrative license—through the collection, analysis, and integration of primary cartographic and demographic resources culled from a variety of archives and digitized library collections, secondary sources drawn from across the social sciences and humanities, period journalism (local, national, and international), and my own photography and ethnographic observation. This methodology reflects historical geography's traditional commitment to vernacular landscapes, which, for Jackson et al. (1979, 6), entails the careful study and appreciation of "an ensemble which is under continuous creation and alteration as much or more from the unconscious processes of daily living as from calculated landscape design," and "a companion of that form of social history which seeks to understand the routine lives of ordinary people." My approach likewise finds resonance with a recent call for "historical geographies of, and for, the present," especially insofar as such endeavors are collaboratively honed for "highlighting memories of struggle and the rearticulation of emancipatory visions" (Van Sant et al. 2020, 169).

In addition to these popular origins and concerns, Part I is also anchored by the hallmark spatial sensitivity of the discipline of geography. "Geography," Van Sant et al. (2020, 169) axiomatically begin their appeal, "is always a product of history." So too, though, does geography influence, structure, and produce history, as this book will show. But as with the potent ambiguities of chilango vernacular, the conceptual, methodological, and practically infinite material variations of space make its elucidation an extremely complex affair. And while my deployment of a spatial framework for this book belongs properly and unexceptionally to this landscape, my investigation of Mexico City's century of urban revolution is guided by relatively straightforward principles of critical spatial analysis, most especially in the traditions of Anglophone cultural and historical geography and those elaborated in Lefebvre's (1991) *The Production of Space*.[10] Put simply, the dialectical approach to the production of space adopted here concedes the complexity, inconstancy, and mutability of the urban world even while attempting to account for the social conditions of its production (and reproduction) and for its persistent influence on the course of daily life. In the widely cited opening to their auspicious *For Space*, Massey (2005, 9) sums up an especially cogent geographical approach to space through three propositions also shared by the present work, which respectively recognize space "as the product of interrelations," "as the sphere of possibility of the existence of multiplicity in the sense contemporaneous plurality," and "as always under construction." Urban space is also saturated with human meaning; both product and productive of social identities, attachments, affinities, aversions, and conflicts.[11] And as the events and processes explored in both Part I and Part II will demonstrate, geography can variously serve as the motivation, mode, and medium of political struggle. As Ong (2011, 23) argues, "Urban exercises are practical and symbolic practices constituting the urban out of varied situated and global connections," which "both depend on and contribute to emergent forms of spatialization that they seek to plot, transform, and achieve."

But while geographers and others across the social sciences and humanities cultivate spatial sensitivity for the purposes of rigorous academic study, residents of Mexico City (and elsewhere) win this attunement, by choice or by force, through their negotiation of the dynamic political cartographies of everyday life. Part I presents only some of the last century's most salient examples of the tensions and conflicts pertaining to urban geography, as seen through rapid urbanization, autoconstruction and informal settlement, patronage and clientelism, land regularization, political organizing, and disaster and recovery efforts, among other processes.[12] Chilangos are viscerally sensitive to the significance of territory to urban politics, grassroots or otherwise; as architect and organizer Sergio González is fond of saying, "somos, porque somos tierra."[13]

The three parts of Chapter 1—the end of the Porfiriato, the Mexican Revolution, and the consolidation of the PRI (especially under Lázaro Cárdenas); the decades of explosive growth and the Mexican Miracle of Import Substitution Industrialization from the 1940s through the 1970s; and the proliferation of crises of the city and of the PRI in the 1980s—challenge discrete treatments of urban geographical and political development by examining their complex interplay and imbrication, both building on and critically diverging from traditional historiography of the city and of national politics in important ways. The chapter begins with General Porfirio Díaz's dictatorial conversion of a rather sleepy former viceroyal seat into a modern capital replete with Western convenience and diversion, though plagued by political corruption and the wounds of a deep and widening inequality in the distribution of the era's unbridled accumulation of resources. When the Mexican Revolution deposed the great don, a period of chaotic government overthrows and rebellions tore asunder much of what Díaz had wrought, until the stabilizing hand of Lázaro Cárdenas fully consolidated power in the party that would become the PRI, reestablishing control of the country's vast regions and delivering on many of even the most radical promises of the Revolutionary constitution of 1917. In the name of the Revolution, the PRI continued to funnel national resources toward the capital, which had legally become like the personal protectorate of the president in 1928. Between 1940 and 1980, ISI policy fueled rural to urban migration at incredible rates, and the city exploded past the boundaries established for the special Federal District, its frayed and distant edges increasingly populated largely by poor migrants housed in self-constructed dwellings only spatially distinct from the tenement-style housing increasingly characteristic of the old center's deteriorating neighborhoods. As the contradictions of these policies set in, tensions within the PRI came to a head. As its factions began to splinter, the victors turned the party toward neoliberal policies that only further devastated the city's poorest areas, as government employment and social subsidies were slashed in line with international austerity mandates brought on by decades of debt-financed expansion and rampant corruption and graft. Increasing authoritarianism, unfortunate geological events, and the perception of abandonment all led to an increasing distance between the PRI and the city's populace, and calls for a more genuine democracy begun in the 1960s reached alarming levels by the close of the 1980s, resulting in the first serious challenge to PRI authority in the 1988 elections.

The fallout from the 1988 elections substantially remade both local and national political geographies over the next several decades. Chapter 2 begins with the political revolutions initiated by presidential hopeful Cuauhtémoc Cárdenas and his breakaway FDN and later PRD, fueled in part by the capital's grassroots Left. The city that had been the seat of the PRI's power had turned against it, and had grown to mammoth proportions with social, political, and ecological problems that lived up to its tenebrific titles far more than the political party often labeled its leviathan ever had. The grassroots Left's efforts to win this democracy would not produce the intended results, however, as the PRI's ultimate defeat in 2000 came not at the hands of the PRD but rather the conservative Partido Acción Nacional (National Action Party, hereafter PAN), whose leadership took full advantage of the electoral reforms begrudgingly passed in the wake of blatant electoral fraud and revelations of a degree of corruption inside the PRI that could no longer be contained. These incremental reforms, however, allowed the city increasing levels of political autonomy, which the PRD both pursued and exploited from the late 1980s onward, building a powerful base of support in the heart of the capital and a robust political machinery many considered eerily reminiscent of the PRI. In concluding Part I, I argue that three conflictual trajectories emerge from this historical geography, surrounding, respectively: 1) the incredible depth but more especially breadth and variety of urban problems legible in the city beginning in the late 1960s and increasingly dire from the 1980s onward; 2) the multifaceted and mutually reinforcing problems of the city's democratic processes, including especially the fractious nature of left-leaning political parties and coalitions, authoritarian tendencies, and clientelist practices inherited from earlier political eras and alliances, and the suppression of genuine democracy by a variety of means; and 3) a tension between the city and the national state, often expressed as a conflict between the parties

in control of each. Contemporary expressions of each of these three trajectories are then explored ethnographically in the three chapters of Part II.

Opening the ethnographic explorations of Part II, Chapter 3 follows a
prominent set of civil society groups and social movements that coalesced
around the idea of the right to the city in the late 1980s and early 1990s.
Using data collected from the archives of NGOs and organizations at the
core of this coalition of forces, along with archived media reports, expert
interviews, and fieldnotes collected during participant observation, I follow this group and its ideas as they partnered with a PRD mayor and several rights-based institutions in the city to produce and promote the Mexico
City Charter for the Right to the City from 2008 to 2010 and the political
uncertainty that followed. In the summer of 2010, the Charter received the
public endorsement of the mayor and the majority of the city's *delegados*
(sub-municipal executives), and subsequently played a prominent role in
several local struggles (exemplified by advocacy for the establishment of
Parque Cuitláhuac in Ixtapalapa and resistance to the elevated expressway
known as the Supervía Poniente). The political significance of the Charter
diminished rather quickly, however, in the face of a new mayoral regime
less than enthusiastic about its principles, and moreover about the political baggage it now carried as an initiative of a previous administration. I
argue that the very dialectical nature which made the concept attractive
to forces focused on simultaneously addressing the city's now legendarily
large and increasingly dire set of ills also formed the basis of its executive
rejection at the hands of the new mayor, who insisted that the city to which
the Charter's demands had been addressed simply no longer existed. Still,
the Charter's promoters held out hope that it might still serve as the onset
of the final stage, as so many had initially demanded, of the city's political emancipation from the national state and the party that controlled it,
and perhaps form the basis of a new urban constitution as a part of this
process. As elaborated in Chapter 5, these hopes were indeed born out in
2017–2018, reviving the promise of the right to the city years after its purported political death.

In the interim, the right to the city languished as the members of its
coalition pursued other issues and agendas. Chapter 4 presents the case
of resistance to a planned redevelopment project in the city's Cuauhtémoc
borough in the fall of 2015, which delivered the powerful PRD a significant political blow. Using ethnographic observations, interviews, and extensive media and social media coverage and debate, I follow the public rollout and eventual defeat of the Corredor Cultural Chapultepec, the first of
a rumored ten megaprojects PRD mayor Miguel Ángel Mancera and his

public/private development czar had planned across the city. The project, which would have converted one of the city's historic avenues into an elevated lineal park designed by the architect son-in-law of billionaire telecom and real estate mogul Carlos Slim, was defeated roughly two-to-one in a public consultation, a decision the mayor unexpectedly decided to honor. I view this process through the lens of postpolitics, which I argue encourages a focus trained on the ways in which opportunities for genuine disagreement are preemptively foreclosed, such that what remains subject to public debate, scrutiny, and participation is only a desiccated set of predetermined options that allow for precious little alteration of the plans of elites. I argue, however, that the narrative often associated with the postpolitical condition is overly totalizing, often missing or ignoring the emergence of politics that fall outside an unnecessarily privileged style of revolutionary transgression of an established police order. Focusing instead on resistance to such orders, I argue for their consideration as postpolitical hegemonies produced within and constrained or enabled by the specificities of time and place. In this case, PRD control over planning and public participation was won in large part by a promise of increasing democratization. The Corredor saga demonstrates the vulnerability of the mayor and his party's hard-won hegemony at precisely this point, necessitating a conceptual reappraisal of the nature of urban postpolitics and suggesting a productive avenue for grassroots movements.

An often-cynical grassroots and resident community enlivened by the unexpected victory of the anti-Corredor forces turned their attention the following spring toward the culmination of the process many hoped the Charter would begin, the restoration of full local democracy to the capital city. Chapter 5 builds on Part I's examination of the radical reorganization of law at the local and national levels known collectively as political reform, which was bent on precisely this end. The chapter begins with a set of ethnographic scenes through which I seek to elucidate the quotidian political climate within which political reform was pursued and enacted in 2016. The product of a much-maligned pact between the PRD mayor and the PRI president (the Pacto por México), these reforms dissolved the Federal District and elevated Mexico City to (roughly) equal status among the Union of Mexican States, and made provisions for the drafting of the city's first constitution, timed so that its conclusion would coincide with the ratification of the country's Revolutionary constitution a century before in 1917. These reforms—which seemed to suddenly arrive at a breakneck pace amid a frenzy of public pronouncements and publicity campaigns that usually contained only partial information about what was after all a sweeping set of changes and processes (some of which were subject to public participation)—were received with a complex mixture of skepticism and excitement by a populace well used to political trickery and manipulation but nevertheless enticed by the idea of genuine reform and the potential of real democratization. To contextualize these often-contradictory sentiments, I explore what I refer to as a revolutionary structure of feeling (following Williams 1977) that continues to surround and, I argue, structure public reception of and reaction to political reform, drawing parallels to memories, feelings, and hopes never fully erased by a century of meandering and sometimes wayward revolution. I organize this analysis around four common sets of meanings or referents for revolution, distilled ethnographically: 1) the Mexican Revolution begun in 1910 and the contest over its legacy and legitimacy; 2) a specifically urban revolution, as experienced demographically and politically in Mexico City during the twentieth century and as expressed through the conflict over the political autonomy of the capital; 3) the city as revolutionary vanguard, from Porfirian efforts to modernize the capital to contemporary struggles over gentrification and *blanqueamiento* (whitening) and the progressive advances of the 2017 constitution; and 4) the notion of revolution as a redemptive process, such that foundered or repressed movements of the city's past are revived and find their promises fulfilled by the revolution now thought underway with the onset of political reform and especially the presidency of Andrés Manuel López Obrador.

In the book's conclusion, I briefly present three lessens on the pace and path of urban change synthesized from my analysis of grassroots politics and Mexico City's urban revolution. First, I argue for a fundamental reconsideration of how urban change is assessed, especially as concerns the success and failures (or life and death) of particular movements, ideas, and interventions. Built around a critical reassessment of Gorz's (1967) "nonreformist reforms," my second lesson suggests the need to expand thinking and practice on the pace of change and the utility and orientation of incremental reforms as a part of a larger revolutionary strategy in the urban age. A third lesson draws on the persistent hope, creativity, and stubborn refusals found in the foregoing chapters. While both urban theory and political realities often present dire dyadic choices, Mexico City's residents have routinely shown the potential of imagining and practicing invasive alternatives, even in the face of seemingly invulnerable opposition. These lessons, learned from a turbulent century of revolutionary urbanism in Mexico City, collectively offer a tempered optimism for grassroots politics on an urbanizing planet.

Part I. Urbanizing Revolution

CHAPTER 1

A Century of Monsters, Machines, and Megaurbanization

The city has become a monster, an urban disaster, a planner's nightmare. For those living in an increasingly blighted urban space, some of Mexico City's most celebrated literary portraits—like Humboldt's description of the crisp mountain air and clear blue skies—now read like sarcastic jokes meant to rub salt on the city's wounds. **Rubén Gallo (2004, 5)**

The revolting monster of the Valley of Anáhuac is not to be trusted; it never has been. Not in the seventeenth century, nor in the twentieth, not today, and not tomorrow. But something about my miserable city has seduced millions.
Mauricio Tenorio-Trillo (2020, 72)

By most estimates, Mexico City is the most populous urban area in the Western Hemisphere, and one of the most populous cities in the world.[1] Though it equally finds easy comparison to North American counterparts (picture the national financial and cultural gravity of New York and the political power of Washington, DC spilled over the sprawling and colorful low-built terrain of Los Angeles) and the booming megacities of the Global South, it also finds a certain distinction in the complex mixture of admiration and derision offered for its description, artfully captured by its most common monikers: *el monstruo* (the monster), *la ciudad de los palacios* (the city of palaces), and *el lago de fuego* (the lake of fire).[2] Across academic and popular portraits of the city, monstrosity figures prominently in descriptions and metaphors for this Mexican Moloch and those who have labored to control it, and the sheer scope and scale of its orbital proportions, whether considered relatively or absolutely, undoubtedly merit such estimations.[3] The problems that stem from this monstrous geography are likewise almost mythically formidable, be they political, economic, social, or environmental in nature. Once the Aztec capital of Tenochtitlán atop Lake Texcoco and its splendorous network of loamy fellows spread across a great valley hemmed in by mountain ranges and volcanic peaks that collide to form its southern terminus, the city has made frightful and wondrous impressions on visitors for some five centuries. Contemporary travelers continue to marvel at the monster, though descriptions now often focus on the immensity of the city's sprawling slums and informal settlements; its nearly unparalleled traffic congestion, whether its chaotic streams of private autos or the variegated patchwork of public and private mass transit networks seemingly in a perpetual state of flux; and the levels and pervasiveness of its pollution.[4] It is a city that displays perhaps better than any other what Holston (2009) calls "the perverse paradox," a situation in which nominal democratization is accompanied by the persistence or even worsening of various social ills, including pervasive violence and socioeconomic inequality.[5] It is in some ways a perverse attraction that continues to draw migrants and visitors to the Mexican capital, just as it has for the last hundred years, whether they are enticed by a fleeting siren song of upward social mobility, the staggering beauty of the valley's landscapes and cultures, or the seductive dance of its powerful political dramas.

Mexico City's path to mythic monstrosity has been enhanced by its pride of place in the history of Mexico, a country in which the constant blending of new- and old-world spiritual, aesthetic, political, and economic trajectories is forever on display in new and exciting ways. A far cry from the novel horror Norton (2003) sees in "feral cities," Mexico City's monstrosity is as old as the treacherous geography provisioned by the eruptions of Popocatépetl and its slumbering neighbors, the earthquakes over which their faults preside, and the shifting silts reluctantly shed by the painstaking retreat of the intermontane lakes upon which the Aztec capital was erected. In the wake of the protracted series of civil wars collectively known as the Mexican Revolution that began in 1910, the capital city entered a new phase of growth that would fundamentally alter and complicate its geographies. The Revolution also birthed another monster, however. The ultimate winners of the war itself and the years of assassinations and intrigues that followed formed in their aftermath a political party and eventually an associated electoral and legislative machine that would rule the country and the city without interruption for the seven decades that ended the millennium. As the city grew beyond reckoning, so too did the party alter its strategies and attempt to maintain its control over the expanding metropolis. In the century that followed the Revolution, the party that bore its name and the city that bore its scars lurched and staggered through their mutually constitutive development, often and increasingly at odds as the contradictions of their monstrous, dyadic maturation began to flare.

This chapter and the next will explore the century or so of urban revolution that produced the two great monsters of twentieth-century Mexico—its capital city and its ruling party. This historical geography will follow similar but more targeted treatments of the city and/or the PRI, such as Eckstein's (1988) consideration of urban development, the urban poor, and the legacy of the Revolution, Davis's (1994) detailed history of the PRI and its relationship with the growing metropolis through the lens of its expanding transit networks, Bruhn's (1997) careful study of the dynamics of the PRI and the breakaway PRD, especially in the crucial 1988 elections, and Gutmann's (2002) assessment of Mexican democracy as seen through an ethnography set in one of the capital city's *colonias populares*.[6] What this chapter and the next will attempt, however, differs from these previous studies both in its emphasis on the changing political geography of the city through the twentieth century and in its consideration of the multifaceted and expanding set of crises faced by national and local governments that resulted from decades of unrestrained growth and the policies and actions of the PRI and its executive heads in particular.[7] In examining the exponential expansion of the city in the post-Revolutionary decades, every attempt will thus be made not only to illustrate the entangled scales of state authority and the everyday geographies with which they attempt to contend, illustrating the fault lines created in each by the other, but also the problems that spin out of this dangerous dance of urban and political development through the twentieth and into the twenty-first century. The historical geography presented here is necessarily selective, and draws attention to both broad trends and a number of particular exemplary moments in an effort to provide a generative introduction to the contemporary political geographies of Mexico City and their tendril connections to the everyday lives of the capital's millions of residents.

Some recent historiography has sought to move beyond narratives and periodizations that twin the catastrophic urban explosion of the twentieth century and the dismaying development of the urban leviathan that became the PRI, which tend to rather too easily cast both as hopelessly grotesque products of development policy and governmental mismanagement variously regarded as misguided or malignant.[8] "In this telling," Vitz (2018, 6) explains, "Mexico City encapsulated the ruling party's bungled governance, emerging as a microcosm of national ruination and disempowerment." Such a telling, however accurate, "occludes a more open-ended history," and runs the significant risk of reifying the dominant chronicle of urban and political disaster, itself often overdetermined by the most putridly florid language. Excavating alternatives to such treatments is a worthy and necessary pursuit. My own purpose in relating the historical geography of this chapter and the next, however, lies in uncovering the ways in which conventional narratives of urban monstrosity—however judiciously derided—continue to structure urban politics in contemporary Mexico City, especially as critical historical junctures (the centennial of the Revolutionary constitution of 1917, for instance) drew near. From presidential campaigns and municipal policy to residential organizing and neighborhood squabbles, popular politics calls upon and manipulates the most hackneyed narratives of urban and political history with startling regularity. "These traditional watersheds are here to stay," Gillingham and Smith (2014, 10) say of such popular partitionings, "in part because they also are embedded in popular memory, products of a nostalgia that invoked (and invokes) Cardenismo as a critique of PRIísmo, and the early PRI as a critique of the later PRI." As Part II of this book will show, both the geographical and political ravages of Mexico City's urban revolution and the contested space of narration that surrounds and suffuses it exert enormous influence on contemporary political dynamics in the most ostentatious and insidious ways alike.

This chapter is organized into three temporal periods: 1) the Porfiriato (when the city was under the control of Porfirio Díaz); the Mexican Revolution, and the consolidation of the PRI in the presidency of Lazaro Cárdenas (1934–1940); 2) the decades of explosive growth and the Mexican Miracle (roughly the 1940s through the 1970s); and 3) the crisis period from 1980 through to the PAN victory in the 2000 national presidential elections. In each, both general trends and particular events and personalities will be highlighted to represent the broader arc of conflict-laden development across this historical geography.

THE PORFIRIATO, THE REVOLUTION, AND THE PARTY[9]

José de la Cruz Porfirio Díaz Mori was first elected to the Mexican presidency in 1877, a post he would relinquish only for a brief four-year period (1880–1884) until he abdicated in early 1911 and fled to exile in Europe. A general who had participated in several major wars, including the War of Reform, Díaz began his political career as a liberal reformer who championed the fight against presidential reelection. Ironically, he would later amend the constitution of 1857 (twice) to ensure the legality of his own hold on power in perpetuity (after initially turning presidential power over to trusted subordinate Manuel del Refugio González Flores after his first term). Given both the enormity of his vision and the degree of control he was able to achieve over the various arms of the Mexican state, military, and even the notoriously resilient and fiercely independent regional *caciques*, his influence on the trajectory of Mexican political economy and the development of the capital city would be difficult to overestimate.[10] Hailed abroad as the great modernizer and the *caudillo* to end *caudillismo*, Díaz brought stability and centralized control to a country whose political culture in the seven decades since its independence had been plagued by civil wars, foreign invasions, multi-term presidents (including Antonio López de Santa Anna Pérez de Lebrón, who sat the presidency some eleven separate times), and a string of presidencies that ended by violence or abdication under threat of violence.[11]

Don Porfirio's reforms and projects pulled together a war-weary and highly regionalized and rural country, and transformed Mexico City into a modern, fashionable capital replete with fancies and splendors fawned upon by a steady flow of foreign visitors and the wealthy elites and political insiders privileged with a seat at the general's table. To secure control of the country's far-flung and highly diverse regions, Díaz recruited a ruthless, rural police force loyal to himself. "By the 1880s," Johns (1997, 68) reports, "the *rurales* had shed enough bad blood to kill most of the brigands and to cause those who thought about taking their place to think again. Their job of pacification done, they became a rural police force that made sure things went Díaz's way." These rurales, along with Mexico City's police force and the military, were pacified by steady promotions and even more by Díaz's highly permissive stance on graft (Johns 1997; Caistor 2000). As Joseph and Buchenau (2013, 22) concede, "Had Díaz not consolidated his particular regime, it is possible, even likely, that the lesser caudillos he (and Juárez) neutralized would have continued to tear the republic to pieces." With political authority thus concentrated and the stability of the regime thus ensured, Díaz pursued an aggressive program of modernization in Mexico's infrastructures and several of its prominent industries. On his watch, Mexico's rail system in particular grew immensely, linking the country's distant ports, mines, and agricultural regions.[12] The regime looked to the cities of Europe for aesthetic and political inspiration in the renovation of the capital, which was swept clean by a loyal police force (but often otherwise left to its vices, provided these didn't offend the *gente decente* [decent or respectable people], in recreational and leisure aesthetics that increasingly converged toward those of the wealthy in the North Atlantic).[13] Praise flowed in from foreign dignitaries and observers—as did foreign capital—as Díaz reestablished the country's financial solvency and prioritized the servicing of sovereign debt.[14] In a familiar theme, the favor of the noisy neighbors to the north and their kindred of the North Atlantic was caustically consequential, and also rather capricious. With the local climate already abuzz with rebellious energy, the same foreign press and *extranjeros* that had sung Díaz's praises for more than two decades turned against him, no longer willing to stomach the increasingly obvious political repression and widespread graft and corruption of his regime, nor the gap between the ostentatious wealth of Don Porfirio's inner circle and the grinding poverty of the country's underclasses. The same *New York Times* that lauded the "master builder" for decades turned the corner on Díaz with notable haste in 1911, labeling him an "old man who dwells largely in the past, and has been kept or has kept himself ignorant of the extent of the disaffection in his country." Duplicity notwithstanding, the *Times* noted the crucial and often overlooked significance of geography to the increasingly untenable contradictions of the regime, the seeds of which were sewn through Díaz's own efforts to modernize the country and its capital city:

> Diaz [*sic*] himself has educated his people for something better; he has built schools and extended free education throughout the country; he has encouraged industry, and secured for Mexico an honorable and important place among the Nations. The natural result is that there is no longer a Porfirist Party to depend upon. The people demand a real republic and the enforcement of the Constitution.

The dissipation of US support further paved the way for the eruption of revolt from among Mexico's disaffected populations, from the dispossessed and otherwise landless peasants who spanned the country to those of privileged birth kept at arm's length by the regime. Amidst growing pressure, Díaz stated his intention to retire from the Mexican presidency in 1910, in an interview that appeared in *Pearson's Magazine* in 1908. Though he declared that his leadership had finally prepared his country for democracy, he ultimately broke this promise, which served as the final straw for his political opponents. In response to his refusal to honor this commitment, Francisco Madero González, who had been plotting his revolution relatively without interference from the US government while exiled in Texas, launched the Mexican Revolution in the late fall of 1910, and within six months Porfirio Díaz had abdicated.[15] The next ten years would witness several periods of open warfare between the major factions of the Revolution and the state, and, in select arenas, the US Army and Navy. These years of war would undo much of the difficult work Díaz had done to integrate Mexico's diverse regions and centralize its powers of governance in the capital city, to garner positive international attention and foreign capital, and to promote significant industrial development. This conflict also brought to the fore a tension between the capital city and the national territory over which its governments attempted to extend their authority. This tension would take a variety of forms over the next century, and would play a significant role in shaping both geographical and political developments as politicians, parties, and citizen movements vied for the upper hand in a series of multi-scalar, socio-spatial conflicts.

A period of relative stability set in with the onset of the Sonoran Dynasty, a series of three presidential terms from 1920 through 1932 (plus a brief interim presidency in 1920) won by generals and politicians from the northern State of Sonora. These included, respectively: Felipe Adolfo de la Huerta Marcor, leader of the Agua Prieta Revolt (the last major overthrow of the Revolutionary period) who served as Interim President for roughly six months in 1920; General Álvaro Obregón Salido, who served as President from 1920–1924; and Plutarco Elías Calles, who served as President from 1924–1928 and exerted great influence on the Mexican government for several years thereafter as the so-called *Jefe Máximo*. The turmoil of Obregón's assassination shortly after his reelection in 1928 brought the government under the unofficial control of Calles, to a greater or lesser degree, for the better part of the next six years (a period known as the Maximato). In 1929, he sought to solidify the Sonorans' control of the state by organizing the Partido Nacional Revolucionario (National Revolutionary Party, hereafter PNR). Through this organization, Calles intended to direct the policies of the executives to serve in his wake, and during his roughly six years as Jefe Máximo largely succeeded. The election of Lázaro Cárdenas del Río in 1934, however, brought an end to the Maximato and initiated a dramatic transformation of the country and the city.

Unlike his predecessors, Cárdenas took great care to forge relationships with even the lowliest of his national patrimony. He began by taking an extended cross-country trek which brought him to every state and territory of Mexico in late 1933 and early 1934.[16] As his victory was assured (with the backing of Calles), this national tour appeared a rather superfluous maneuver. This gesture, however, was a necessary first in a series of brilliant political strokes in early Cárdenismo. To do away with or at least mitigate the overbearing Calles and his already well-established political machinery in the capital, Cárdenas required a massive base of reliable support. This he largely found in the country's peasants and laborers, to whom he made major concessions through his celebrated agrarian reforms and permissive treatment of labor strikes, respectively. It would still take Cárdenas two years to fully depose the Jefe Máximo and his cronies, but when a crucial public misstep (a miscalculated insult in the press regarding Cárdenas's treatment of labor strikes in the capital) provided him the pretext he required to remove Calles to exile in California, substantial support throughout the country insulated his regime from the threat of revolt and brought much of Calles's machinery under his control at the head of the party he would rechristen the Partido de la Revolución Mexicana (Party of the Mexican Revolution, hereafter PRM) in 1938.[17] In this new guise, the party that had begun as the vehicle of the late Sonoran Dynasty and the personal fiefdom of the Jefe Máximo took on an entirely different arrangement. His popular bases of support, workers and peasants, had their mass organizations brought directly inside the party as distinct sectors, in addition to the military and state employees. These four sectors were expected to deal directly with the president and his representatives, a move which sought both to further curb the influence of caciques and to enhance the power and control of the party's executive head. At the close of the *sexenio* of Manuel Ávila Camacho, Cárdenas's handpicked successor and longtime loyal subordinate, the party would again be renamed, achieving its final nominal form as the Institutional Revolutionary Party (PRI).[18] In addition to his massive land redistributions (largely in the form of communally held *ejido* grants to peasant and indigenous communities throughout the country) and his extensive nationalization program, Cárdenas rearranged Mexican politics such that the country would no longer be dominated by one person, as it had been under the de jure and de facto rule of Calles.[19] Rather, Cárdenas's leadership would subordinate Mexican politics to a single party, the PRI, for the remainder of the century.

Cárdenas's sexenio also witnessed a profoundly consequential reorganization of the capital, which would politically subordinate the Federal District under the national executive.[20] This move would have lingering implications of growing significance even until the present, and paradoxically created a situation in which a revolution nominally based in an effort to establish democracy had done precisely the opposite at its very center, such that some sixty years later Ward (1989, 308) would conclude that "few places in the democratic world have less local democracy than Mexico City." Through a 1928 amendment to the Revolutionary Constitution of 1917, the existing *municipios* of the Distrito Federal (initially established in 1824) were reorganized as *delegaciones*, further legally distinguishing the city from the states of the Mexican Union and placing the city's legislative agenda in the hands of the national congress.[21] The city's chief executive (*regente*)—indeed, much of the city's government—was from then on installed by appointment (by the president of the Republic), rather than election.[22] The Regente was primarily tasked with keeping peace in the capital and otherwise managing the carefully guarded and highly coveted electorate housed there, which del Carmen Moreno Carranco (2008, following Ward 2004) argues produced political appointees whose tenures were of two types, based either on adeptness and effective governance (as in the sixteen-year reign of Ernesto Uruchurtu) or personal loyalty to the president (as in the case of Manuel Camacho Solís, who worked closely with Salinas to stymy early attempts at Federal District political reforms in the late 1980s and early 1990s). At the national level, the presidency dominated the national legislative process by proposing the legislation to be debated by the houses of congress, and also proposed constitutional amendments, which then required a two-thirds congressional majority and ratification by seventeen of thirty-two (as of 2016) states. This incredible degree of executive authority, bolstered by the PRI's longtime control of the national congress and its domination of the federal bureaucracy, Edmonds-Poli and Shirk (2009, 78) argue, "made legislators a mere rubber stamp for executive legislation, budgetary approval, and even constitutional amendments," which numbered "over 400" between 1917 and 2000.[23] To say nothing of the rest of the country, this arrangement submitted virtually all functions of government in the Federal District to the direct or indirect control of the federal executive, and thus placed the city's affairs under the personal purview of nearly seven decades of PRI presidents. As Perló Cohen (1993, 129) argues, "Para bien y para mal, la Ciudad de México se convirtió en la ciudad presidencial" (For good and bad, Mexico City became the presidential city).

The city was also beginning to grow at previously unknown rates as modernization efforts drew rural populations to burgeoning urban industrial centers. In 1900, the Federal District contained some 542,000 inhabitants. At the outbreak of the Revolution ten years on, that number had grown to roughly 730,000, and by the end of Cárdenas's sexenio it had exploded to between 1.6 and 1.8 million, an increase of over 300 percent.[24] The geographical footprint of its urbanized area, however, largely remained well within the bounds of the Federal District (see Figures 1.1, 1.2), and some

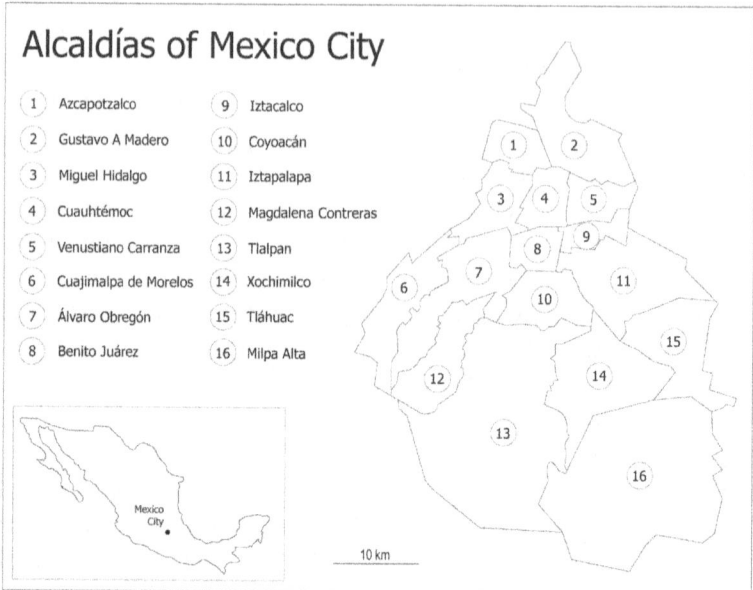

FIGURE 1.1. The alcaldías of Mexico City (formerly the delegaciones of the Federal District). Cartography by author.

cardinal elements of the city's political and symbolic architecture remained in place from earlier stages of development, including especially the Zócalo (the city's central plaza), the city's administrative offices and the buildings that housed the legislature and other arms of the federal government, and significant relics from the Viceroyalty of New Spain and Tenochtitlán before it, such as the famed Metropolitan Cathedral and some of the ruins of the Aztec Templo Mayor.[25] Grand projects begun before the Revolution would be unsubtly liberated from their moorings in service to Porfirian progress and repurposed to ply their splendors for the Revolution. Many are now as engrained in the mental maps of residents (to say nothing of the promotional materials of the city's boosters) as the Zócalo itself, including El Palacio de Bellas Artes (begun in 1904), El Monumento a la Independencia (generally referred to as El Ángel de la Independencia, completed in 1910), and El Monumento a la Revolución (begun in 1910).[26]

By the 1930s, the city's geography lent itself easily to the modeling that had become a mainstay of a burgeoning urban studies in the academy of the neighboring United States. Mexico City conformed more than most to general descriptions of urban spatial form in Latin America, such as those later offered by Portes and Walton (1976) and Griffin and Ford (1980). As Garza

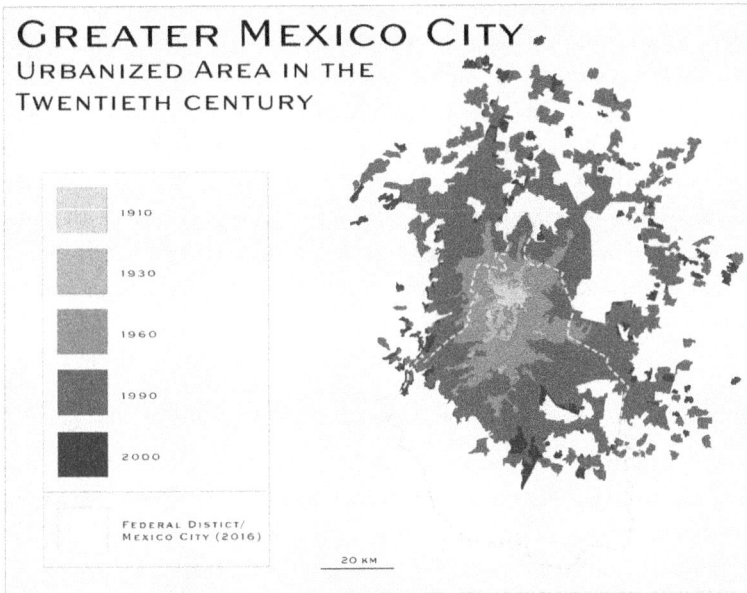

GREATER MEXICO CITY.
URBANIZED AREA IN THE
TWENTIETH CENTURY

1910

1930

1960

1990

2000

FEDERAL DISTRICT/
MEXICO CITY (2016)

20 KM

FIGURE 1.2. The greater Mexico City urbanized area through the twentieth century. Data adapted from Ward (1990) and Connelly (2003). Cartography by author.

Merodio (2006) cautions, however, the city's adherence to international (or even regional) models and terminology should not be overstated, especially as concerns industrial development within the city, which would not reach recognizable maturity for several decades after the Porfiriato.[27] Nevertheless, the Porfirian vision of urban development had enormous impacts on the city's aesthetics and geographies, initiating tendencies that would only accelerate in the decades that followed. A series of increasingly well integrated tramway and railway networks and stations allowed an early exodus of elite families toward the city's western and southern edges, along with a general spatial expansion of the city's urbanized area to accommodate a rapidly growing population. New colonias began developing around the historic center, including working class (such as Guerrero and Santa María la Ribera) and wealthy residential neighborhoods (such as Cuauhtémoc and Juárez).[28] Governmental decisions about whether and how to infrastructurally incorporate these growing communities followed a logic that would predominate during the exponential growth of the decades to follow. "In effect," Lear (2001, 29) explains, "the government only provided services if they thought they could recuperate costs from taxes on the resulting properties, a condition only the wealthier *colonias* could meet." Though ravaged by the Revolution (which also sent a large number of Porfirian elites into exile), the city would resume this pattern of growth and modernization from around 1920. As PRI regimes from Cárdenas on worked to produce and sustain the period of national economic growth known as the Mexican Miracle, the city's growth would transcend the mythic boundary between quantitative difference and qualitative change through the establishment and settlement of new colonias at rapidly increasing distances and densities, the breakneck pace of industrialization and modernization, and the exertion and eventual contestation of political control under a maturing "urban leviathan" (Davis 1994).

URBAN EXPLOSION AND THE MEXICAN MIRACLE (1940–1980)[29]

Because of the immense changes wrought in and on the city during the four following decades, Mexico City continued to serve as a primary exemplar for social scientists seeking to contend with urban Latin America, though on completely different terms. Within the Federal District, rapid population growth more than doubled the population, which exceeded nine million by 1980. Far more dramatic changes, however, were seen at the scale of the metropolitan area. In 1940, the Federal District's population (roughly 1.65 million) constituted roughly 94 percent of the population of the metropolitan area (roughly 1.76 million). By 1980, however, the population of the metropolitan area had grown to over 14.4 million, only 64 percent (roughly 9.2 million) of whom lived in the Federal District. In other words, during these four decades Mexico City experienced incredible peripheral growth, far exceeding that of its central areas. In this, el monstruo figured as only the most populous example of a regional and even global trend, as

FIGURE 1.3. Iztapalapa, just before the Via Crucis (Good Friday) celebration. Photography by author. **FIGURE 1.4.** Iztapalapa, near the Cerro de la Estrella (Hill of the Star), just before the Via Crucis celebration. Photography by author.

FIGURE 1.5. Looking roughly east from the Cerro de la Estrella in Iztapalapa, just after the Via Crucis celebration. Photography by author. **FIGURE 1.6.** The corner of Avenida 5 de Mayo and Avenida Ferrocarriles Nacionales, outside of Metro Refinería, Azcapotzalco, looking roughly northeast. Photography by author.

FIGURE 1.7. The Village of Texcoco at the edge of the Mexico City region (in the neighboring Estado de México), looking roughly east down Nezahualcoyótl Street from the second floor of Cantina Las Palomas, with the Cerro de las Promesas (Hill of Promises) visible in the distance. Photography by author. **FIGURE 1.8.** Behind a market in Texcoco. Photography by author.

FIGURE 1.9. Texcoco, seen from atop the ruins of the Los Melones archaeological site, with friend, roommate, and occasional research assistant Ignacio in the foreground. Photography by author

urbanists, policy-makers, philanthropists, and other interested observers turned their eyes to the plight of the peripheral poor in exploding cities in Latin America and Asia in particular. This peripheral settlement came in many forms across distinct geographies, from the slopes and hillsides of Iztapalapa (see Figures 1.3, 1.4, and 1.5) and Ciudad Nezahualcóyotl to the flatter northern areas around Azcapotzalco (see Figure 1.6), and the varied terrain of swamps and rolling hills that surround the colonial village of Texcoco (see Figures 1.7, 1.8, and 1.9), near the site of the city's now twice-cancelled new international airport.[30] These new areas were largely settled informally, through means that included everything from unauthorized squatting and occupation to shaky purchase or rental agreements that may or may not find the protection of law should proverbial push come to shove.[31] The resultant sprawl of neighborhoods—referred to since 1945 as *colonias proletarias*, but largely known as *colonias populares* in contemporary parlance—are "characterised by cheaply acquired land, inadequate infrastructure, and self-help dwelling construction . . . often developed on agricultural land" (Lombard 2014, 19). Within the space of a few decades, in other words, the city of palaces had spilled outward in rapid fashion, and was now ringed by a tumultuous patchwork of peripheral settlements that spread well into the hills on the edges of the Valley of Mexico.

Life in Mexico City's expanding colonias populares unfolded in conditions and along patterns that exemplified urban marginality in extreme relief.[32] Autoconstruction dominated the housing options in the settlements that sprung up in this period, and they were (and in some cases remain) chronically underserved by public utilities such as potable water, waste removal, and electricity, to say nothing of safe and reliable transportation. But as Griffin and Ford argue, these newly developed areas served as sites of semi-permanent or permanent settlement and community-building, and many of these areas were soon integrated within the economic and social circuits of the city.[33] As with the colonias constructed during the late Porfiriato, incorporation, regularization, and service delivery proceeded unevenly based on the political and fiscal calculations of city government.[34] Clientelist practices flourished as the party's strategy of appeasement by incorporation saw a swelling bureaucracy trade services for support (in only the most general of terms). Permissive oversight, opportunities for patronage, longstanding (but uneven) property regularization programs, and varying levels of public investment in housing and urban services created a variety of opportunities for gradual upgrading in the city's expanding constellation of colonias populares, as Duhau (2014) argues. Such a

progressive habitat, though afflicted with all the various ills of urban informality, housed the labor that powered the PRI's booming industrialization,
devolved many of the costs of rapid urbanization to individuals and communities of rural migrants, and created reliable paths to political incorporation—and, as Perló Cohen (1979) argues, control—by establishing durable
patronage networks across the city's expanding geographies.[35]

Residential densities across the city grew and shrank unevenly during
this period of immense growth, but in patterns that display several interrelated trends. Densities increased rapidly in the poorest areas of the metro
region, especially in the colonias populares at the northern and eastern
edges of the Federal District and several municipalities of the neighboring Estado de México, and decreased pronouncedly in the core neighborhoods of the central city, especially in the Centro Histórico.[36] But as Ward
(1990) convincingly illustrates, the pronounced dip in densities around the
city center—which in Connelly's (1988) two-dimensional graphic depic-
tion yields an image not unlike that of a volcano, as densities first balloon
throughout the city then suddenly drop just in the center beginning just
before 1960, only to then rapidly rise and slowly taper off toward the edges
of the metropolitan area—and the rise in densities moving outward from
there, illustrate two distinctly classed movements. On the one hand, rural-
to-urban migration was fueling incredible growth and densification at the
peripheries, while on the other hand the historic center was being abandoned by wealthy families who had already begun to move out and settle
along the historic Paseo de la Reforma, the prototypical example of what
Griffin and Ford (1980, 407) describe as "a commercial spine surrounded
by an elite residential sector," and the neighborhoods beyond, eventually
including the area of highly elite residences tucked into the hills west of the
Bosque de Chapultepec known collectively as las Lomas and the colonia of
Polanco, and also areas in the south of the Federal District long home to
the estates of elite families, as in the neighborhoods of San Ángel, Jardínes
del Pedregal, and Coyoacán.[37] These two logics of growth produced distinct spatial patterns, with the wealthy tending to push the city's expansion
through exclusive residences toward the west and the south, and poorer residents and informal settlements driving densely packed expansion toward
the north and east. In 1954, a city measure aimed at curbing rapid density
increase, what Ward (1990, 40) describes as "a ban imposed . . . upon the
authorisation of low-income sub-divisions in the DF" was, Ward claims, a
failure on two fronts. Ward argues that not only was the ban not effective
in curbing density in the Federal District—owing both to its timing, which

came far too late for some of the neighborhoods where it would have been most effective, and lax enforcement by city officials and outright disregard by landlords and tenants and fueled by intense competition for space— but also and moreover fueled rapid growth in the neighboring municipalities of the Estado de México, especially Nuacalpan and Nezahualcóyotl, grossly exacerbating metropolitan expansion and effectively pushing the problem of irregular and high-density settlement out of the Federal District's jurisdiction.

All of this enormous growth was accompanied by and to a large degree enabled and encouraged by concomitant national and local economic growth, sustained over a period of roughly three decades in what came to be called the Mexican Miracle. Prompted initially by the nationalizations of Cárdenas—who finally put into practice some of the most radical appropriations provisions of the 1917 constitution—the next several PRI regimes continued to nationalize and/or ensure majority-Mexican ownership (Mexicanization) of selected industries and otherwise follow a protectionist path, in line with the regional trend of economic thought and practice known broadly as import-substitution industrialization (hereafter ISI).[38] The PRI–state also engaged in broad interventionist policies including through the national development bank, Nacional Financiera, which served to heavily capitalize selected industries, and the imposition of prohibitive tariffs on consumer goods.[39] As the descriptor *miracle* suggests, Mexican ISI policies produced some three decades of sustained economic expansion.[40] This growth in the national economy was disproportionately advantageous to Mexico City, which, as the seat of government and the primary locus of early industrialization in this period, was in a position to garner major investment and the fruits of patronage (Ward 1990; Davis 1994). The public sector expanded with the national economy, as state expenditures ballooned with state ownership and backing of industry, an expansion that chiefly benefitted the growing middle classes housed in the metropolitan area. As Alexander (2016) argues, this middle-class professionalization of the city was an intentional pursuit of the PRI, and was quickly reflected in the growth in influence of the CNOP sector of the party (at the expense of both the CNC and the CTM, the industrial labor and peasant organizations within the party, respectively).[41] Importantly, the benefits of growth also accrued unequally across class lines more broadly, both in the city and at the national level. While the middle classes (especially at the top end) expanded and gained a larger share of the national income, inequality also rose markedly overall, affecting the poorest Mexicans most especially, and, according to Ward (1990, 7), "Income distribution remained one of the most unequal of all Latin American nations."[42] As in the rapid economic growth and urban development under Díaz, then, that of the Miracle decades sewed afresh the seeds of class contradiction in the rapidly growing metropolis. Other contradictions of ISI policies viewed at the macro scale also came into focus as the 1960s wound down, including the paradoxical imbalance of trade that resulted from the need to import capital-intensive inputs for primary commodity production in particular, such as in the vast oil industry now under the control of the behemoth national petroleum company PEMEX (Petróleos Mexicanos), and the increasingly troubling debts required to sustain the PRI's course of development and pursuit of a sustainable domestic consumer market.[43] But while the tools of macroeconomics could be readily employed to combat these financial contradictions (such as currency devaluations in the late 1940s, mid-1950s, and mid-1970s, or later neoliberal reforms to social spending and national enterprise in the 1980s and 1990s), the urban spatial contradictions of the miracle decades were more far more acute and, for the party, ultimately more dangerous.

To tend the city's growth, promote social and political stability, and ensure predictable electoral support, PRI presidents carefully selected their mayoral subordinates. Perhaps the most ambitious of these mayors was so-called Iron Regent Ernesto Uruchurtu Peralta who occupied the office from 1952 through 1965 and whose regime exemplifies many of the complex problems already beginning to confront the ruling party from within its most prized domain. Though known for his ability to effectively carry out the wishes of the presidents he served, Davis (2009) argues that Uruchurtu also attempted to cut a middle path between contradictory PRI impulses during the Miracle decades, especially during the administration of Gustavo Díaz Ordaz Bolaños (1964–1970). Though he fostered growth through redevelopment and beautification projects in the city, Uruchurtu attempted to limit the scope of the damage to the properties and broader spatial arrangements of the city's growing middle classes and government employees, overseeing impressive new middle-class housing developments such as the Nonoalco-Tlatelolco housing estate, and even protecting the massive Tepito market (and city landmark) La Lagunilla. He also publicly opposed plans for the city's first Metro system and the havoc they promised to wreak on the historic center in particular. The Metro system—begun in 1965 only after Uruchurtu's removal, based, according to Davis (1994, 2009), on his stubborn opposition—would ultimately only further exacerbate the city's peripheral growth patterns as publicly subsidized and speedy transit spread to far-flung neighborhoods in the ensuing decades. The subsequent three mayoral administrations, none of which completed a full sexenio, were each plagued by their own issues, the most glaring and significant of which will be discussed below. The administration of Carlos Hank González brought a return to rapacious redevelopment with seemingly little regard for affected communities or political repercussions.[44] While no overall pattern explains the governance style nor the particular political choices of the Regente regimes of this period, Uruchurtu's dilemma of managing the on-the-ground contradictions of the PRI's national economic and political agenda is instructive. Indeed, as Davis's (1994) thorough analysis of PRI dynamics at the interface of national and local regimes clearly illustrates, the PRI was never quite the leviathan it has often been made out to be, as mayoral and presidential regimes and the competing vertically integrated sectors of the PRI frequently came into intense conflict, pulling at the very fabric of the party's corporatist structure and making its hold on power ever-more tenuous, especially as its decades of co-optation continued to bring new constituencies into the fold.[45] Especially as the popular sector of the party grew and benefitted from ISI policy, and as the city's peripheral development swelled the unincorporated ranks of the urban poor, the PRI's leadership began to lose the ability to effectively generate consensus, and its executives were forced to find new and, in some cases, disturbing ways to bend the state and the populace to their will. These tensions were nowhere more apparent than in the gathering conflict between the national state and its sometimes-unruly subordinates in the Federal District.

An additional demographic factor that came to play a major political role in this developing tension was the massive and rapid bottom-end growth of the city's population pyramid. Ward (1990) argues that though the main driver of metropolitan growth during the early part of the Miracle was undoubtedly rural-to-urban migration, it was soon displaced by natural increase.[46] Unfortunately for PRI leaders, a generation of children born during the first decades of ISI were reaching the age of political participation just as the magic of the Miracle began to peter, especially at the city's periphery. In 1968, the country's most storied political movement was formed among these youths, whose complaints ranged from a growing competition for scarce jobs to the unsatisfactory provision of urban infrastructures in peripheral developments, and, perhaps above all, the PRI–state's lack of genuine democracy and what they viewed as its unjustified detainment of political prisoners. Owing to its cultural and political significance,[47] the history of what came to be called the student movement—including its origins, composition, and demands—is a matter of considerable and ongoing debate.[48] Whatever the specific weight of grievances represented by the student movement, they included both economic and political claims, though neither were of a particularly radical nature. Paz (1985, 233) characterized their demands as reformist, writing a year later in 1969, "They were not expecting a radical change, but they did expect greater flexibility and a return to the tradition of the Mexican Revolution, a tradition that was never dogmatic and that was very sensitive to changes in popular feeling." The student movement exemplifies a change in common perceptions regarding the PRI state and its hardening since the Revolution and its consolidation under the Sonoran Dynasty and the early experiments of the PNR and PRM. If the conflict between Regente Uruchurtu and President Díaz Ordaz had demonstrated a certain dangerous factionalism within the leviathan, the student movement was an attempt to protest the party's increasingly authoritarian practices, the legitimacy of its *dedazo*, its presidencialismo, and other elements of its approach to political rule, as well as its ability to continue to deliver on the promises of the Revolution.[49] For Bolívar Echeverría (2019), the movement represented far more than student grievances, and was far more connected to the demands of a burgeoning urban society than is often appreciated.[50] It was an urban political awakening, and a significant challenge to the party's legitimacy.

Though energy was reportedly beginning to dwindle by the end of September, the movement was immortalized by the shocking massacre perpetuated by the state on the afternoon of October 2nd. Ten days before the opening ceremony of the 1968 Olympic Games held in Mexico City, protesters gathered in the Plaza de las Tres Culturas, which is bordered by the Nonoalco-Tlatelolco housing estate.[51] Though this history continues to be disputed by the Mexican government in particular and many of the details remain murky, it is abundantly clear that several factions of the Mexican military were sent by President Díaz Ordaz to surround the plaza in the late afternoon. A general scholarly consensus (which follows a popular consensus established within days of the events) places the blame for what transpired next at the feet of the president and some subordinates, including Luis Echeverría Álvarez, then serving as secretary of the interior, who would follow Díaz Ordaz as president two years later. A small group of trained sharpshooters fired on the crowd, prompting immediate chaos and the onset of a more generalized violence on the part of the military surrounding the plaza.[52] Though government officials initially grossly underreported the number of victims, careful reporting places the number of dead persons around three hundred, with some hundreds or perhaps thousands more injured.[53] The episode, which came to be known as the Tlatelolco Massacre, effectively ended the student movement, at least as it had been.[54]

The student movement and the killings at Tlatelolco continue to reverberate across Mexico, and especially in the capital. As Tamayo (2015) and others have convincingly demonstrated, the movement had profound and lasting influence on the orientation of the city's grassroots militancies, ideologies, and political cultures. The massacre brought forth from the state the very thing the movement had sought to expose and roll back, the PRI government's authoritarian practices and its suppression of grassroots political mobilization. This event only deepened the legitimacy crisis of the PRI–state, though it would take some years for this to find an oppositionary expression with real force. In its wake, as Crane (2015) argues, generations of Mexican activists would come to view the state as a statically repressive entity, virtually unchanged by the march of time and the course of politics, at least so long as the PRI was at the helm. As Paz (1985) argued, Tlatelolco demonstrated the party's betrayal of the Mexican Revolution, in so far as it was threatened by demands for fairly minor reforms and demonstrated a violent opposition to change, an intransigence that many considered anathema to the identity it claimed as the institutionalization and perpetuation of the Revolution's principles. And though the timing of the student movement's demonstrations was intended to garner international support for their claims—or at least to offer a highly public critique of the PRI–state's own claims about Mexico's progress and modernity—the Olympic Games took place without incident (ensured by a heavy military presence), and the world hardly seemed to take notice of the murder of (at least) several hundred students and sympathizers at the hands of a government simultaneously touting its democratic legitimacy.

Tlatelolco thus represents a watershed moment in the relationship between the PRI and the Mexican populace, especially in the capital city. A shockingly similar mass killing of students would take place less than three years later on June 10, 1971. Known as the Halconazo (the hawk strike) because of its perpetration by the shadowy elite military group known as los Halcones, this incident served to further galvanize both the growing anti-PRI sentiments and seemingly the party's own resolve to stay in control of the capital. Together, these events irreparably tarnished the name of President Echeverría, whose administration also conducted a large-scale dirty war campaign against rebellious elements in the city and the countryside through his sexenio (1970–1976) and the PRI, as successive regimes continued in the use of torture, extrajudicial executions, and disappearances over the next several decades. In 2015, public records offering some details of this campaign were put beyond the reach of the public, locked in a secure collection in the National Archives housed in the former Lucumberri Prison—the

Black Palace, where dissidents are rumored to have been tortured during the Porfiriato—in the Centro Histórico. Few of the likely grossly underrepresentative number of crimes the Mexican state has since acknowledged took place during this dirty war have ever been punished.[55]

Despite these troubles, the PRI continued to win elections unchallenged throughout the 1970s and well into the 1980s, even as increasing numbers of chilangos lost faith in the party and its leadership, and even as the party's internal tensions began to blossom into their own crises. Economic growth, the party's lone salvation beyond its growing repression, continued at acceptable rates through most of the decade, though fluctuations became more pronounced as the above-noted contradictions of Mexican ISI policy began to come into focus and macroeconomic policy pivoted toward exports in the 1970s (Ball and Connelly 1987). Inflation continued to be a problem, born largely of the ironic trade imbalances brought about through rapid industrialization, which brought further currency devaluations and sharpened the pains of the city's growing inequality. To keep the Miracle going, Presidents Echeverría and his successor José Guillermo Abel López Portillo y Pacheco were forced to borrow ever larger sums, pumping cash into the inflated state sector and the many heavily capitalized industries now dependent on state support. From 1970 to 1983, total foreign debt increased ninefold, from just under US$10 billion to over US$90 billion, also an increase of over 60 percent as a share of GDP (up to over 80 percent from around 20 percent in 1970) (Moreno-Brid and Ros 2009). The following decade would bring these fiscal chickens home to roost.

The four decades from 1940 to 1980 brought incredible changes to Mexico City and to the PRI. The party found its final form after its consolidation under Cárdenas in the 1930s, and pursued yet further nationalizations of industry and intensified and sustained a suite of import-substitution industrialization that produced a fantastic rate of economic growth, which came to be known as the Mexican Miracle. The growth of industry and state-sector employment drew millions of migrants to the city's neighborhoods and more especially to its peri-urban edges, which were largely developed through organized land invasions, piecemeal squatting, and other informal processes. While these populations remained deeply intertwined within the circuits of the urban and national economies, their conditions of life mirrored those of other marginalized populations on the peripheries of the many other exploding cities of Latin America, where ISI policy also favored cities at the expense of the provinces. The exacerbation of inequality in Mexico City was only the most obvious of the costs of this development

program, however, and it became clear through the PRI–state's treatment
of the generally reformist student movement of 1968 and the intensification of its ongoing dirty war against dissidents and rebellious elements
throughout the country during this period that the increasingly authoritarian party faced a significant legitimacy crisis. The party and the state
also faced the pressures of its exploding foreign debt, itself an attempt to
conjure the means to keep the Miracle in effect and to service the state's
still-growing commitments to a city whose growth was finally starting to
slow and a national economy whose contradictions had almost caught up
to the magicians.

THE TURBULENT 1980S: THE EMBATTLED LEVIATHAN AND THE RESURGENT URBAN LEFT

As of 1980, the city's spatial and demographic explosion had begun to quell
(see Figure 1.2), owing to several factors whose influence can be difficult to
disentangle. The physical geography of the metropolitan area certainly presented some limits, as there is only so much available and inhabitable space
in the valley. This is further complicated by the relative spatial configuration of employment opportunities, accessible and affordable transit options,
and fluctuations in the metropolitan area's land markets, themselves influenced by idiosyncratic tenure and title regularization and the regulation of
plot and building subdivisions, among many other factors. Thus is the state
implicated in what otherwise appears a passively received condition. Macroeconomic factors also played a considerable role. While the Echeverría
and López Portillo administrations had been able to maintain economic
growth (albeit with increased volatility), the costs of this growth required
a new course to be charted in 1982, one less favorable to urban industrialization and metropolitan expansion. The city's demographic transition followed a fairly standard modernization trajectory, as birth rates eventually
fell dramatically in the years following rapid industrialization.[56] By 1980,
the metropolitan area's overall rate of population growth was less than half
of what it had been a decade earlier, bringing an end to the period of largely
unrestrained urbanization.

López Portillo's debt spending in particular had been buoyed by the discovery of large new oil reserves during the early part of his tenure, though
the revenues they ultimately produced failed to match the administration's
early predictions, at least partly due to fluctuations in global oil markets
and a consequent drop in price in 1981. The next PRI president to receive
the dedazo was Miguel de la Madrid Hurtado, whose background and

approach to economic policy represent both a transformation of the party's economic orientation and a growing tension within the party regarding precisely this change and its implications. De la Madrid came from the growing faction of the PRI known as the *técnicos* (technocrats), many trained abroad at prestigious universities (de la Madrid held a graduate degree from Harvard's Kennedy School) in such disciplines as economics or public policy, who increasingly found themselves at odds with the orientation of the so-called *políticos* (who continued to find advancement through the traditional channels within the party and its organizations). Among other differences in their trajectories and political philosophies, técnicos followed other contemporary Latin American leaders in finding great appeal and policy inspiration in neoliberal economic theory. As O'Toole (2003, 270) puts it, "such an intellectual climate had by the late 1970s contributed to a polarisation of perspectives on the country's economic future between advocates of neo-Keynesian expansionism, legitimised by nationalism, and monetarist orthodoxy, the legitimation for which derived above all from a more positivist belief in progress resulting from the application of technical economic expertise." In de la Madrid's case, some of the far-reaching reforms that flowed from this general orientation happened also to be imposed from without, as his government required assistance from the IMF in particular to pull its way out of its billowing fiscal catastrophe, an embarrassment deeply felt by the party whose economic Miracle had been the envy of the region. As Eckstein (1988, 221) explains the situation as of late summer 1982, "by international standards, Mexico was bankrupt and technically in default, a circumstance unknown internationally since the 1920s. Until Mexico's crisis, international creditors had assumed that a petroleum exporter was immune to such problems." To fix the disaster, de la Madrid turned to deregulation and privatization. On his watch, the protectionist infrastructure and ISI heritage of the PRI for the previous forty years and more was cut to ribbons. In two years alone (between 1985 and 1987), the number of imports subject to licensure was cut by over 90 percent, and rates that had been as high as 100 percent were capped at a maximum of 40 percent (Eckstein 1988). Also under de la Madrid, Mexico joined the General Agreement on Tariffs and Trade (GATT) in 1986, and would continue to pursue free trade agreements and the expansion of production in the special economic zone along the US-Mexico border, where *maquiladora* manufacturing intensified through the 1980s and into the 1990s. Social spending was slashed as austerity measures were instituted in the name of restoring the nation's credibility, and the bloated state bureaucracy was chopped down to a fraction of its pre-crisis size.[57] State-owned enterprises were sold off, though the state retained an interest in what it designated strategic areas, namely "petroleum and basic petro-chemicals, telecommunications, electrical power generation, nuclear energy, railroads, and banking" (Haber et al. 2008, 69). According to Haber et al. (2008, 69; following Valdés Uganda 1994), "There were 1,155 state-owned firms, public trusts, and decentralized agencies at the beginning of the de la Madrid administration; there were just 412 by its end," and *La Jornada* (2009) reports that the de la Madrid sexenio witnessed the fire-sale-figure of 63 percent of public enterprises sold off to private interests. De la Madrid's government also engaged in widespread wage suppression through the manipulation of the minimum wage, which only further served to decrease the already falling purchasing power of the poor and increase inequality generally.[58]

Carlos Salinas de Gortari, another prominent técnico then serving as de la Madrid's Secretary of Budget and Programming, sought to position the austerity measures within the PRI's claim to the principles of the Mexican Revolution. In a public ceremony on the November 20 Revolution Day of 1983, Sheppard (2011, 513) reports, "Salinas differentiated between what he called 'reactionary austerity' and the revolutionary variety, arguing that revolutionary austerity seeks the reordering of the economy, creates the conditions for Mexico to overcome structural difficulties and deficiencies, and prepares the country for the future." Thus the timing of the 1982 debt crisis was highly opportune for one group, as it presented a critical juncture through which the burgeoning técnico faction of the PRI was able to forcefully assert itself.[59] Though there were many who preferred to view the crisis as the result of blunders on the part of López Portillo (whose nationalization of Mexico's banks just before leaving office in 1982—amidst the gathering crisis—was accompanied by massive rallies in the name of renewed nationalism and reassertion of the PRI's commitment to revolutionary social and economic policy, though this nationalist fervor did little to insulate him from criticism in the ensuing years), de la Madrid and his advisors attempted to use the onset of the crisis to propagate a passive revolution to neoliberal economic policy.[60] Even within the PRI, the técnicos' assumption of control and subsequent tectonic political-economic shifts alienated many of the party's leaders and exacerbated increasingly untenable factional tension.

De la Madrid's economic policy also failed to duplicate the successes of the Miracle years, and served to intensify several contradictions that the debt crisis had allowed to surface. First, social and economic inequality in Mexico intensified through the 1980s, spurred on by the massive cuts in social spending and state sector employment.[61] Employment opportunities in newly privatized industries seem to have offered less stability and economic security, and purchasing power declined steadily through the 1980s and into the 1990s, even among those employed in manufacturing, which showed promising growth in nominal wages. And as de la Madrid sought to pivot resources away from the capital through decentralization—in an effort to both democratize *estados* and municipios throughout the country by devolving certain budgetary and financial authority from the central state, and to some degree to spread the frustrations and embarrassments of the debt crisis and subsequent reforms across these entities—the city's poor and marginalized residents were left further behind. The informal sector of the city's economy absorbed many of the unemployed and underemployed as it continued to expand during this period.[62] Though by definition such activity eludes reliable calculation, studies such as that of Benería (1989) demonstrate that the informalization of much of Mexico City's economic activity in the 1980s disproportionately affected the most vulnerable activities and members of the labor force, particularly women. The generally lackluster growth of the economy, the continuing debt problems, and the ravages of neoliberal austerity programs conspired to see the 1980s later described as the lost decade.[63]

While the timing of the 1982 debt crisis had been fairly opportune with respect to the machinations of the PRI's técnicos, it could not have been worse for the city's poorest residents, especially those living in the vecindades of the city's central neighborhoods. Three years later, at 7:19 a.m. on September 19, 1985, an 8.1 magnitude earthquake rocked the city, killing some ten thousand people, especially in and around the Centro Histórico. The following evening, another quake struck with near equal force.[64] Owing to its underlying geology, the valley shakes with some frequency, and earthquake alarms are a common auditory assault in many neighborhoods to this day.[65] The earthquake of 1985, however, was singular both in its destructive force and more importantly in its social and political implications. Coming as it did at the tail-end of decades of rapid population growth and geographical expansion, and in the midst of de la Madrid's neoliberal reorganization of the PRI–state, the earthquake struck the city at its most vulnerable. Makeshift construction fueled by decades of self-help housing practices folded and collapsed without resistance, as did many of the historic center's centuries-old structures, palacios and vecindades alike. Though accounts again vary, the earthquake certainly completely destroyed some four hundred buildings, many of them crowded apartment buildings and subdivided structures home to poor families. Gilbert (1998, 139) claims that some ninety thousand homes were damaged, thirty thousand of which had to be demolished in the aftermath.[66] The PRI response to this catastrophe was perceived by many victims and residents generally as confused, slow, and ineffective.[67] Though at the national level there were emergency plans in place for such an event, de la Madrid was reportedly hesitant to use them because they were not specifically designed for the capital, and more importantly because they relied on the military, which de la Madrid feared could swell popular support for a military long held in check by the PRI (and whose dedicated vertical sector within the party had been dissolved decades earlier) (Camacho de Schmidt and Schmidt 1995). Instead, the official response lagged, with significant delays even on time-sensitive issues such as the search for survivors and the restoration of basic services.[68] As many have argued, the PRI state's response to the 1985 earthquake represented another damning piece of evidence for many of the city's residents that their government simply could not be trusted, and that the party had betrayed both the Revolution and its constituents. Residents of many central neighborhoods refused to wait for an inattentive and ineffectual state, however. Groups formed across the affected areas to perform necessary functions, such as search and rescue, medical assistance, coordination of shelter, and protection from opportunistic looting and other threats to person and property. In the ensuing months and years, many of these groups transformed into robust social movements and alternative political forces. This paroxysmal popular organizing is now commonly heralded as "the birth of 'civil society' in Mexico, a spontaneous means by which informed citizens would assert their own destinies and place popular demands upon the state" (Smith 1989, 397). Azuela (2007, 161) argues that the phrase itself (*sociedad civil*) jumped scale with these events, as the philosophical penchant of a small group of high-minded leftists "rapidly became part of the political vocabulary of the whole country." Notable among this groundswell of organizing were several of the constituent groups that together make up the Movimiento Urbano Popular (Urban Popular Movement, hereafter MUP), and the Asamblea de Barrios, whose comical mascot Superbarrio and the zany public feats he performed in no way diminish the seriousness of the group's political influence.[69] As Davis (1994, 281) points out, many such neighborhood groups had already begun to emerge prior to 1985, encouraged by the limited measures implemented to maintain the middle- and lower-class character of some neighborhoods close to the historic center in particular and responding to "the evident neglect of popular demands" even before the earthquake.[70] The earthquake and the PRI response galvanized these existing efforts into new solidarities born of necessity amid unfathomable loss and suffering among the multitudinous chilangos who would come to be known as the earthquake generation(s).[71] For Tenorio-Trillo (2020, 70), the existential necessity to "amizarse—to reinforce oneself, to make oneself solid" became a resilient ethos forged in the inferno of the aftermath, with profound political implications palpable in everyday life: "Skill at reinforcing things, and at mistrusting—those were the gifts won by the earthquake generations." Their emergence within particular neighborhoods ensured localized, grassroots political agendas and resilient territorial identities for many of these groups for decades to come.

Such groups were often posed a choice in the aftermath of the earthquake: continue to organize locally and avoid the uncertain and potentially dangerous world of national party politics, or join with opposition parties in an attempt to oust the PRI through the standard procedures of Mexico's nominal electoral democracy.[72] Some of those that chose the latter course found an ally in PRI leader Cuauhtémoc Cárdenas Solórzano, son of beloved General and President Lázaro Cárdenas. Along with other PRI leaders deeply unsatisfied with the party's turn toward neoliberal policy—which they viewed as a desertion of the party's and the Revolution's social and economic principles—Cárdenas founded the Corriente Democrático (Democratic Current) in 1986. The group intended to apply pressure within the PRI, steering it back toward earlier economic and social policies and influencing the selection of the party's next presidential candidate. They argued, writes Bruhn (1997, 83), that "shrinking the public sector as a solution to our financial woes amounts to cutting off the head of a man suffering from a headache." When the Corriente failed to effect significant change to either the economic trajectory of the party or its internal democratization—owing in large part to the party being largely controlled by the new generation of technocrats who lined up behind Carlos Salinas de Gortari as their nominee for the 1988 presidential elections and sought to isolate and discredit members of the Corriente—Cárdenas and other lifetime *priistas* broke away from the party entirely and mounted an impressive opposition campaign. Cárdenas's support was organized through a *frente*, a coalition of forces that offer formal support and official representation to a single candidate, in this case the Frente Democrática Nacional (FDN). "Twelve separate organizations signed the Common Platform of the National Democratic Front," Bruhn (1997, 136) reports, "three political parties with legal registry, five parties without legal registry, and four popular/civic movements." The inclusion of each of these groups held distinct advantages. Cárdenas required a legally registered party (most of which had checkered histories with the PRI) to support his electoral admission for the 1988 elections, for example, while urban social movements in Mexico City bolstered his credibility and ultimately his overwhelming support in the capital. His campaign also did well in certain rural areas (especially in the states of Baja California, México, Morelos, and his home state of Michoacán), and the legendary status of his father ensured that rural and urban communities alike greeted the younger Cárdenas with an affection and respect that impressed even those skeptics who doubted his ability to raise an electoral army large enough to slay the embattled leviathan. And despite the disadvantages that came with a dispersed coalition taking on a powerful political machine, his support swelled in the months just before the election, and a nervous tension gripped the country as international observers prepared the onlooking world for the arrival of "real" democracy in the world's eleventh-most-populous country and, despite the downturn of the preceding years, still its sixteenth-largest economy.[73]

Election day came on July 6, and early indications showed strong support for Cárdenas and the FDN.[74] As the votes were being counted that evening, however, a crash of the system was announced—allegedly a problem with the electronic vote-counting machines employed for the first time during this election.[75] All counts were halted on the order of President de la Madrid, and in the hours that followed Salinas was announced as the winner without any official figures. Three days later, official tabulations gave Salinas a mere simple majority with 50.36 percent of the vote, with Cárdenas receiving 31.12 percent and Manuel Clouthier del Rincón (of the conservative PAN) receiving 17.07 percent. Many were outraged at what seemed an obvious fraud, especially in the capital.[76] Parliamentary procedure, however, allowed the PRI to elicit a measure of cooperation from the legislature in legitimizing the election results. Each legislature was at that time required to undertake an initial process of *autocalificación* (self-qualification) of their own election (and, in years of coinciding presidential elections, of the presidential election) before they can be formally seated.[77] Therefore, as the smaller parties that officially or unofficially supported the FDN sought to solidify their own gains in the legislature, they sacrificed their last best chance to prevent the seating of Salinas, and dashed the presidential hopes of a leftist coalition that represented the first serious electoral challenge to the PRI in nearly six decades.[78] In a rhetorical vein reminiscent of his attempt to square the circle of placing backbreaking austerity at the heart of Revolutionary ideals nearly five years earlier, Salinas applauded his manipulated election as "unquestionable and evident," and as heralding the arrival of genuine electoral plurality, and, therefore, renewed democratic legitimacy.[79]

Over the next six years, Salinas would deepen the PRI's commitment to the new course it had set under de la Madrid, privatizing state interests, cutting social expenditures, and encouraging foreign direct investment and free trade. Arguably his administration's crowning achievement was the trilateral North American Free Trade Agreement (hereafter NAFTA), a bargain that set off the Zapatista rebellion in the south of the country.[80] The pro-growth policies of the Salinas administration, however, failed to yield the promised returns, and his sexenio was deeply marred by major scandals and high-profile crimes, along with increasing poverty and intensifying inequality throughout the country as state subsidies disappeared and what remained of the country's social infrastructures was dismantled, and as NAFTA fully opened Mexico's markets to US consumer and agricultural products.

CONCLUSION

This chapter has begun an historical geography of Mexico City, focusing on its political development and that of the party that ruled it for the majority of the twentieth century. Beginning with the immense changes wrought on the landscape and in the halls of state by Porfirio Díaz, the capital city became a favored site for investment, especially in industry and international commerce. The city's population grew as Don Porfirio sought to centralize state authority, quash regional rebelliousness, and project an image of economic and social modernity that would show world powers that Mexico was ready to join their ranks. When the contradictions unleashed by his economic plans—most notably both the rampant inequality at work in the modernizing Mexican economy and the political repression by which this economic growth was ensured—conspired with personal blunders and gathering regional unrest to force his ouster in 1911, a chaotic era characterized by the promulgation of reforms accompanied by frequent political intrigues and assassinations set in for the capital city and would not be completely quelled until the presidency of Lázaro Cárdenas in the mid-1930s. At the close of this presidency, the capital and the country over which its authority extended had been completely reorganized. The capital city had entirely lost its local democracy, having been remade into a Federal District under the jurisdiction of the president of the Republic. In the ensuing decades, the ruling party embarked on a period of import-substitution industrialization that swelled the population of the capital to an enormous size and pushed its spatial frontiers well beyond the bounds of the Federal

District as its newest residents struggled to craft a livable habitat from even the riskiest hardscrabble parcels at the furthest peri-urban remove. As the party maneuvered to maintain the economic Miracle, it faced crises on several fronts, and its troubling responses to these challenges opened up yet further fissures and stoked gathering popular unrest in the city. A party whose power was based in a national electorate and national machinery was increasingly finding itself at odds with an urban populace with less and less to lose, and without much hope remaining in the official bearers of the Revolution's mantel. Mexico City's rapacious but episodic urban revolution, that is, had fostered the growth of an urban political consciousness, especially in the wake of mass mobilizations in 1968 and 1985. As Mexico's long twentieth century neared its end, its two great monsters were thus increasingly entangled in dangerous conflict.

CHAPTER 2

Crisis, Conflict, and
Cárdenista Revolution

This saturation of events produced an increased hunger for the company of narrative in the form of news and commentary, a feeling of living in historical times, and a heightened sense of both the past and future, as experience ceased to be relevant for meeting the challenges of the present and past expectations were strewn about like so many wrecks made uninhabitable by an uncontrollable wave of events. This heightened historical sensibility is implied by the invocation of the term *crisis*: an interruption of the pertinence of past expectations and experience in the present. **Claudio Lomnitz (2003, 134)**

Es el éxito de la reforma política que impulsó el PRD; es la primera vez en la historia del país que hay un sistema electoral confiable, que habrá un gobierno autónomo en la capital; es el fin de la hegemonía del PRI y, por primera vez, comenzará a funcionar el sistema de equilibrios concebido en la Constitución de 1917. **Porfirio Muñoz Ledo (quoted in *La Jornada* 1997)**

Bitter defeat in the tragicomically suspect 1988 election was a severe blow to Cárdenas's leftist coalition, including urban social movements, ordinary *capitalinos*, and perhaps most especially the *priistas* who had hitched their wagons to his political star and followed him out of the party. The return to the Revolution's principles that Cárdenistas hoped and organized for throughout the 1980s would have to be found by another path. The party that Cárdenas formed from the dashed hopes and loose remnants of the FDN, the Partido de la Revolución Democrática (PRD) would come to govern the capital city in opposition for the next several decades, and its relationship to urban social movements, democracy, and revolution form much of what follows in subsequent chapters. National politics would take a far different electoral course through successive PRI regimes marred by corruption, the long awaited democratic transition of 2000, two PAN regimes remembered mostly for an enormous spike in violence and a perceived lack of control, and the return of the PRI with a fresh new face and the promise of yet further democratic reform. In the capital, these national sexenios and their municipal counterparts witnessed a host of socio-spatial changes, from the first victories in the long struggle over a suite of measures known collectively as political reform and the consolidation of municipal power under leftist leaders new and old to the reshuffling of aesthetic codes and investment cartographies that would set the stage for wholesale neighborhood change in the decades to follow. As with the previous chapter, this chapter will provide an overview of these spatial and temporal transformations with a particular eye to their simultaneously political and geographical development. The chapter concludes Part II of the book by elaborating three trajectories of conflict that emerge from this historical geography as enduring legacies that form part of the structure of contemporary Mexico City's everyday political geographies.

REFORM AND REORIENTATION

Despite the promising performance of the FDN in 1988, the PRD failed to remain competitive in presidential elections. In 1994, Cárdenas received less than 17 percent of the national vote, despite increasing dissatisfaction with Salinas and the PRI's neoliberal orientation. Instead, opposition support began to flow toward to the right-leaning PAN, whose candidate Diego Fernández de Cevallos Ramos won just under 26 percent of the vote.[1] As a consequence of nascent electoral plurality in the 1988 legislative elections, Salinas was compelled to make significant overtures to the PAN, especially in the Chamber of Deputies.[2] As the neoliberal turn of the PRI made the conservative PAN a more natural ally, noninterventionist economic concessions seemed anything but concessionary. Broad electoral reform, the true price of PAN cooperation, was a far more concerning expense for the PRI. Over party objections, Salinas relinquished PRI control over much of the electoral process in order to stay the threats that had emerged in the election and the chronic popular disillusionment they portended.

The breadth of the electoral reforms of the late 1980s and early 1990s would be difficult to overstate. Under Salinas and his successor Ernesto Zedillo Ponce de León, the electoral process received an overhaul rivaled in scope and size only by the still unfolding neoliberal reforms begun in earnest under de la Madrid.[3] Arguably the most significant of the 1988 reforms saw the Instituto Federal Electoral (Federal Electoral Institute, hereafter IFE) established as an independent and unaffiliated body, ostensibly insulating future national elections from manipulation by the PRI.[4] Other changes touched "almost every conceivable aspect of electoral organization: a new voter card; the creation of reasonable campaign spending limits; and finally, generous public funding and access to mass media for all parties, not just the PRI" (Langston, 2017, 68). Zedillo's later reforms were born of their own pressures. The last presidential gasp of the PRI, Zedillo became the PRI nominee only after the assassination of presumptive nominee Luis Donaldo Colosio Murrieta in Tijuana in March of 1994 left the party scrambling.[5] The early years of his presidency witnessed the collapse of the Peso (for which Zedillo requested and received US and IMF assistance, which did not help his popularity), the ongoing EZLN uprising, and several notable political scandals.[6] Amid these troubles, a suffering public image, and widespread speculation on the demise of the party, Zedillo agreed to and promoted the 1996 reforms to demonstrate his—and the PRI's—commitment to meaningful democratization and an end to political corruption. In painfully ironic fashion, concessions offered to maintain a fragile national hegemony became the ground upon which a new and ultimately more effective resistance to PRI dominance was constructed by the PAN through the 1990s. In 2000, PAN candidate and longtime Coca-Cola executive Vicente Fox Quesada handily defeated PRI candidate Francisco Labastida Ochoa and by then perennial PRD candidate Cárdenas (who again received just under 17 percent of the vote), heralding the arrival of "real" electoral democracy to Mexico at long last and closing nearly a century of revolution, institutional and otherwise.[7]

The increasing pressures of plurality after 1988 forced the PRI to focus on electability far more than it had in previous generations, and provoked a concomitant reassertion of territory as a political concern.[8] Geography found different expressions in the strategies and fates of the three parties, however. Through the 1990s, the PAN's national strategy was to parlay PRI concessions into newly contestable legislative and gubernatorial seats across the country, gaining ground in piecemeal fashion on its way to capturing the national executive in 2000. The party garnered its largest share of the national vote in 1994, portending the transition to come. The PRI, however, relied on traditional tools of incorporation in its reconquest of the capital city that had turned against it in 1988. In what Foweraker and Landman (1995, 203) consider a "successful and long-term campaign strategy" focusing on urban voters, the PRI reincorporated many of the defunct FDN and fledgling PRD's supporters in the early years of the 1990s, especially in the

capital, and rode the promise of Salinas's NAFTA success and poverty alle-
viation through PRONASOL to a strong showing in 1994 despite the scandals noted above. Though successful, for a time, at the national level, the
PRI would see its share of a democratizing GDF erode as the 1990s wore
on. By contrast, the PRD's territorial orientation, while a clear failure at
the national level, first exploited and then effectively remade the political
geographies of the capital city, especially after 1996. Alongside the city's
most powerful social movements and other grassroots political forces, the
PRD pushed a conciliatory PRI for dramatic reforms in the Federal District, much as the PAN did at the national level. Though forged by a series
of failed national presidential conquests, Cárdenas's party would take on
an increasingly urban character and focus as it pursued and navigated the
changing landscapes of an increasingly competitive Mexico City.

In the local lexicon, the protracted struggle for the political liberation of
the DF from federal (largely executive) domination and the escalating series
of hard-won changes are often collectively referred to as political reform.[9]
Calls for the capital's autonomy took on a heightened intensity during the
1980s, illustrating an increasingly dangerous political cleavage between the
interwoven geographical scales of city and country. The cascading tragedies
of *la crisis* served to transform what had been a simmering tension into formal demands for urban autonomy.[10] It would take several decades of incremental reforms for the city to wrest the institutions of governance from
the national state, in a slow but steady process in which democratization, a
significant concept at the national and local levels, was deeply intertwined
with the political emancipation of the capital. Until 1997, the Federal District was firmly under the control of the Mexican president, after the 1928
reorganization of the modern Distrito Federal (along with the appointed
position of Regente), as discussed in Chapter 1. In 1988, the Federal District was granted an Assembly of Representatives, though for the next six
years this body was able only to make recommendations. A further round
of amendments gave it a measure of legislative authority in 1993, though
even then this body was not directly elected until a yet further round of
reforms in 1996, when it became the Asamblea Legislativa del Distrito Federal (Legislative Assembly of the Federal District, hereafter ALDF).[11] Arguably the most historic and consequential of these reforms made provisions
for the direct election of a newly rechristened jefe de gobierno for the Federal District. In their first mayoral elections, newly enfranchised chilangos
elected PRD–founder Cárdenas to the post.[12] Thus, while the national electorate expressed its fateful dissatisfaction with the PRI in a rightward drift,
the capital city spoke its inaugural democratic homily in the language of

the Left. The PRD would win each of the next several elections, losing only to AMLO's breakaway Morena party in 2018.

The city's landscapes and geographies were likewise transformed by the power struggles and political intrigues, legal reforms, and economic fluctuations of the 1990s. Political territory was drastically reshuffled as PRI patronage shifted to the PRD, simultaneously reflecting a radical break at one level—with a new party increasingly garnering the political loyalty of neighborhoods and industries that had long belonged to the PRI—and continuation—as many such networks reportedly remained largely unaltered in personnel (who simply switched party affiliations) and tactics—at another. Austerity and the fabled fertility factors of the demographic transition conspired to curb the monstrous growth of previous decades before the city reached the most startling and apocalyptic predictions of the global prophets of population, but the swelling of the city's spatial footprint had changed its relative internal geometries.[13] Neighborhoods that once marked the city's furthest edges were now considered central, and haphazard uneven development was continuously smoothed over and haltingly regularized by the crude and unpredictable alchemy of public and private resources.

The historic center and its surrounding colonias also began a gradual process of transformation, especially as residents, the state, and, more consequentially, real estate capital sifted through the carnage of the 1985 earthquake and its aftermath. Government programs such as Renovación Habitacional Popular (Popular Housing Renovation, jointly funded by the World Bank) rehoused some 78,000 of the *damnificados* displaced by the earthquake in new housing units and tore down scores of the city's damaged vecindades.[14] Desertion of damaged homes and commercial properties nevertheless fueled both further physical deterioration and a growing general perception of these neighborhoods as unsafe that endures to the present.[15] The widespread destruction and wholesale devaluation of these central neighborhoods would also pave the way, in future years, for sweeping neighborhood transformation of the sort often dumped at the conceptual doorstep of a purportedly global gentrification.[16]

THE DEMOCRATIC TRANSITION

After 2000, the defeated PRI was forced to take a nominally junior position in its vague and sometimes tenuous alliance with Fox's conservative PAN.[17] But where the PRI had attempted to reforge its relations with the capital by making historic democratic concessions and vowing to purge itself of political corruption and widespread narco influence, the PAN turned its attention outward. Fox is the grandson of immigrants from the US state of Ohio and the Basque country of Spain who grew up in the Mexican countryside and was famously ridiculed for his folksy and colloquial language. As the standard bearer of a renewed Mexican democracy and a more politically palatable face for continuing neoliberal reform, Fox courted stronger ties with US president George W. Bush, an initially fruitful endeavor derailed by the terrorist attacks of September 11, 2001. Stymied by chance in this major international initiative, Fox's presidency was also troubled at home by crises that would likely have meant little to previous PRI regimes, such as the violent clashes over and ultimately cessation of the aforementioned efforts to construct a new international airport outside the capital near the village of Texcoco, and the country's worsening narco-trafficking problems. Felipe de Jesús Calderón Hinojosa, Fox's PAN successor, is best known for his disastrous attempt to find a military solution to drug trafficking, especially (but far from exclusively) in the north of the country.[18] Both presidencies also failed to produce the kind of economic returns their constituencies had been promised after two decades of chronic instability under the PRI, amazingly perhaps the lone unforgivable sin amidst the cesspool of violence and disarray that marked the PAN's only two terms at the helm of the Mexican executive.[19]

Violence wore different faces in the democratic transition's many transformations of the capital city. Despite the ebbing of the twentieth century's rapid population tide short of the most catastrophic predictions, the city's overtaxed infrastructures and environments had already consummated many of its most apocalyptic prophesies of pollution, corruption, crime, social destitution, and physical decay. While many chilangos maintained what I have elsewhere (2019) called a hopeful rejection of perceived realities that allowed for a forging ahead amidst such difficulties, city government adopted a more aggressive posture. Under the PRD and particularly under AMLO—who succeeded the brief and deeply embarrassing tenure of Rosario Robles—the GDF accelerated the urban renewal efforts of the 1990s in the Centro Histórico and its surrounding colonias, often working hand-in-glove with real estate interests to promote an aesthetic and demographic overhaul of the city's oldest districts.[20] Unsurprisingly, such clean-up and restoration programs targeted street vendors and others toiling and living under informal conditions. In addition to rounds of formal expulsions of such ambulantes, AMLO's government (backed by funding from Carlos Slim, among others) contracted former New York City mayor Rudy Giuliani's security consulting firm, Giuliani Partners, to advise the GDF on its plans to lower crime and establish reliable control over troubled sectors. Marcelo Ebrard Casaubón, then serving as AMLO's Secretario de Seguridad Pública, oversaw the implementation of several of the proposed changes, most of which followed textbook tenets of the broken windows policing Giuliani and Police Chief William Bratton had branded on a 1990s New York City cast somewhere between Bret Easton Ellis and Spike Lee. Some police powers and priorities were reorganized, and new techniques were introduced in service to a focus on improving the city's aesthetics and reputation.[21]

To curb peripheral growth and encourage the planned revitalization of central neighborhoods, the GDF introduced the controversial Bando 2 program in 2002. This initiative restricted new housing permits to the original four delegaciones of the Federal District—Benito Juárez, Cuauhtémoc, Miguel Hidalgo, and Venustiano Carranza—and encouraged both higher-density housing development and rising prices in these most central areas of the city. As Delgadillo (2016, 1170, original emphasis) argues, the changing demographic (toward smaller family units, and younger and wealthier residents) and property (toward projects with both dramatically more verticality and larger unit footprints) profiles of these areas in the first decade of the new millennium amounted to a gentrification "copromoted [*sic*] by the state and the market," of which "*exclusionary displacement* seems to be the current and widespread expression." Incited by the concerted efforts of a purportedly leftist and opportunistically entrepreneurial GDF and a property-hungry nouveau riche itself spoon-fed by decades of neoliberal dismantling and fire selling of public assets, neighborhood transformation swept from Centro down the spine of the Paseo de la Reforma toward Polanco and through sleepy and picturesque colonias like Condesa, Roma, and Cuauhtémoc in the ensuing years, refacing the capital for presentation to a burgeoning coterie of international suitors eager for its reentry to the polite society of millennial global cities. Somewhere in the process, residents began to find themselves on the outside looking in; no longer recognizing their neighbors nor their neighborhoods. And though the party of Cárdenas remained in power, even some of its ardent local supporters began to murmur of betrayal by their leaders as prized urban patrimony masqueraded as hiply refurbished baroque playgrounds and found its way into the hands of the dreaded real estate cartels.[22]

For many contemporary residents, perhaps especially those who report or betray a certain disillusionment with the PRD in recent years, the party's redevelopment priorities and relationships with private development interests have been highly suspect from the outset.[23] While decrying neoliberal policy and ideology at the national level, PRD mayors have presided over a qualitative shift in the character of corporate investment in Mexico City's landscapes that many consider profoundly neoliberal, with projects and initiatives often pursued through public-private partnerships and in concert or consultation with foreign firms.[24]

Like Cárdenas before him, AMLO would resign as jefe in July of 2005, in order to prepare for the presidential campaign of 2006. The GDF was handed over to Alejandro Encinas Rodríguez, an original PRI defector and trusted Cárdenista then serving as AMLO's Secretario de Gobierno. Both tarnished and buoyed by accusations and investigations stemming from his tenure as mayor and a complicated relationship to a rapidly changing capital city, AMLO nevertheless remained a front-runner in what became an exceptionally tight race against Calderón in July of 2006.[25] IFE initially counted Calderón the winner by a razor-thin margin of 0.58 percent, or 243,934 votes, though both AMLO and Calderón had already publicly claimed victory. In what was already becoming a familiar pattern, AMLO cried foul. Citing electoral irregularities nationwide—including illegal use of funds by Fox, voter intimidation and coercion, and electronic manipulation of early electoral returns—AMLO filed suite with the Tribunal Electoral del Poder de la Federación (Electoral Tribunal of the Federal Judiciary, hereafter TEPJF), challenging the validity of the initial count and demanding a nationwide vote-by-vote recount.[26] The TEPJF conceded a much smaller measure, which did little to alter the final tally and handed Calderón a certified victory on September 5.[27] With undeterred conviction, AMLO and his followers convened frequent demonstrations both before and after the official results, culminating in the massive Convención Nacional Democrática (National Democratic Convention, hereafter CND). Held in the Zócalo on September 16, the convention was reported to have registered some 1,025,724 delegates.[28] By a show-of-hands vote, the assembly declared AMLO "presidente legítimo de México" (legitimate president of Mexico) in full confidence of a stolen July election, and declared that he would form a legitimate government ten days before Calderón's inauguration on November 20.[29] Ensconcing his efforts in the canon of Mexican democracy alongside Revolutionary "hero of democracy" Francisco Madero, AMLO announced an ambitious reform agenda for his shadow government.[30] Dramatic public posturing and increasingly divisive squabbling in the ensuing weeks and months combined with domestic and international signals of support for Calderón, however, to dampen the political impact of AMLO's post-election coalition, which began to publicly hemorrhage allies.[31]

Despite another bitter national defeat and embarrassing rifts in the upper echelons of the party (including between AMLO and Cárdenas), the PRD continued to amass greater control of the Federal District during the first decade of the new millennium; both propelling and benefitting from yet further reforms aimed at ever-greater democratization at both the municipal and national scales. While AMLO struggled to maintain the attention of the national populace, newly elected Jefe Marcelo Ebrard Casaubón set about the work of quietly distancing his administration from that of his increasingly incendiary mentor, despite their reasonably consistent municipal agendas. The rescue of Centro and the beautification or redevelopment of other favored sites around the city continued, even over the objections of local communities and grassroots organizations. Public safety and security remained a core element of these efforts, including the continuing intensification of policing aimed at quality of life offenses and zero tolerance enforcement along the lines proposed by Giuliani. Enacting these priorities resulted in the mass displacement of many informal activities and artisans, with drastic consequences (as intended) to the aesthetic character of Centro and other targeted plazas and neighborhoods across the city.[32] Notwithstanding the tensions caused by securitization and redevelopment projects, Ebrard cultivated a favorable rapport with much of the grassroots Left in the city during his sexenio, and esteem for Ebrard seemed to grow even stronger some time after his departure from the *Jefetura*.[33] The PRD also had strong showings in ALDF elections throughout the first decade of the millennium, especially at the expense of the PRI, which likely accounted for some of the shift toward a reportedly more collaborative executive posture toward the ALDF after AMLO's abdication.[34] The increasing control this afforded the PRD, and the uses to which this power was put, is the subject of Chapter 4.

REDUX AND REVOLUTION

After a down showing across the board in 2009, AMLO and the PRD attempted to occupy the national spotlight again in the run up to the 2012 presidential contest. The violence and ineptitude of Calderón's drug war had tallied a gruesome cost in human life, social fabric, and political confidence throughout the country, and the PAN was made to suffer for these and other sins. Many in the capital hoped the celestial alignments favored a leftist alternative at long last. But, again like his by-then-estranged PRD forebear Cárdenas, AMLO's second run at the presidency was less successful than the first. Though theories abound to explain AMLO's defeat by more than seven points to a resurgent PRI, few are as satisfying as the simplest and that most commonly heard in the city: that, despite its many faults, at least the PRI could deliver, knew how to govern, and could realize and exercise control. The PRI had also learned from its mistakes in the early cycles of democratic opening during the 1990s. Focusing on electability, they ran the young and handsome Enrique Peña Nieto, former governor of the Estado de México basted by duodecimally latent arms of the PRI political and media machine in the fairytale glow of Revolutionary statesmanship and the romance of a recent union with popular *telenovela* star Angélica Rivera dripping with Hollywood contrivance. Instead of the lofty ideals and extravagant promises of the leftist populist, Mexican voters opted for stability, beauty, and a return to some semblance of control. After another round of vigorous but ill-fated public declarations of electoral fraud, the capital's "messiah" was forced to spend another sexenio in the wings.[35]

For some observers, the 2012 elections represented a regression or "interruption" (Tuckman 2012a) of Mexico's democratization, while for others the peaceful transition of power illustrated the robustness of the gains made since the election of Fox in 2000. As Ackerman (2014)—who cautioned against the apathy encouraged by the myth of a democratic transition—explains, one enduring legacy of the PAN years was the loss of centralized power under a strong PRI president and the consequent empowerment of regional PRI leadership. This, Ackerman argued, forced Peña Nieto into the uncomfortable and unfamiliar position of needing allies in order to effectively govern. In addition to regionally powerful members of his own party, Peña Nieto immediately sought to build bridges outside the PRI, including in the capital city.

One such ally was the DF's new jefe de gobierno, the PRD's Miguel Ángel Mancera Espinosa, who garnered an historic margin of victory in the 2012 election. Though Mancera would use this mandate to dubious effect, as explored in Chapter 4, it doubtless provided a measure of insulation for some unpopular maneuvers at the outset of his term, including a partnership with Peña Nieto. Brought together by Peña Nieto's need for allies, Mancera's pursuit of further political reform in the capital, and the temporal coincidence of their victories, the new executives of the resurgent PRI and the once rebellious PRD built a close working relationship early in their tenures. Cooperation between the two was quickly consummated by their joint participation in Peña Nieto's *Pacto por México* (Pact for Mexico), an agreement between leading figures in the PRI, PAN, and PRD signed in a December 2 ceremony held at Chapultepec Castle. The agreement secured cooperation on Peña Nieto's proposed educational, telecommunications, and federal finance reforms (Herrera and Urrutia 2012). This alliance was an immediate affront to many leftists and other anti-priistas, and seems to have cost Mancera the confidence of many in the PRD, and certainly that of much of Mexico City civil society. In the months and years that followed, a widespread perception of Mancera's centrism and coziness with the severely unpopular Peña Nieto would only deepen this distrust with both Mancera and the PRD more generally. Popular feelings of disillusionment and betrayal by the PRD fed rebellious inclinations. In the wake of the elections and with the backing of many PRD stalwarts from bottom to top, AMLO broke with the PRD and converted the Movimiento Regéneracion Nacional (Movement for National Regeneration, hereafter Morena) into a formal political party, citing the Pacto as evidence of a hopelessly compromised PRD.[36] As the PRD had done to the PRI, Morena would in subsequent years begin to consume the popular base of the PRD in the capital, converting local leaders, grassroots organizations, and other networks of support. As the chapters that follow will explore, conflicts playing out at several scales would finally conspire to deliver AMLO the presidency in 2018, as well as bringing an end to PRD rule of the city's executive in the person of longtime AMLO loyalist Claudia Sheinbaum Pardo.

What could Mancera and the PRD have weighed worthy of such an alliance as that solidified by the Pacto por México? For one thing, it established a center-Left partnership that helped Mancera keep control of the capital. The PRD was already under threat from Morena, and was also looking toward new challenges from the resurgent PRI. Thus, while the Pacto alienated many of the party's leftists, many were already following AMLO out, stage right. An altogether different and more significant spoil would become clear only four years later, however, as this alliance put political reform back on the national agenda. In various guises, political reform had been a signature project for the city's elected mayors since Cárdenas first sat the post, and Mancera's alliance with Peña Nieto held out the promise of a longstanding and lofty dream: the abolition of the Federal District. The substance of this most ambitious end of political reform and the politics that surrounded its implementation are the subject of Chapter 5.

RECONSIDERING MONSTROSITY

On the eve of the 2012 elections, the *Guardian* published historian Alan Knight's critical review of *Guardian* journalist Jo Tuckman's (2012a) *Mexico: Democracy Interrupted*.[37] Knight concludes his review by arguing, "even if the PRI returns to power, Mexican democracy—with all its many and familiar imperfections—will probably survive, uninterrupted; but so will the country's intractable social, economic and security problems." This assessment, also highly applicable to Mexico City, concisely conveys several interrelated arguments that emerge from the preceding chapters and their treatment of Mexico City's changing political geographies. First, as Knight's rather hopeful assessment of the state of democracy after the 2000 transition suggests, many of the hard-won changes to the political system were robust and likely to remain, including political plurality in local and national elections, the return of local democracy to the capital city, and sweeping electoral reforms. The lasting impact of these changes on Mexican governance at the local and national levels, however, was less clear, and is a theme to be taken up in later chapters. Second, Knight's attempt to temper the swell of negative responses—from disappointment to rage— likely to accompany the PRI's return to presidential power is informed by a more complex understanding and treatment of the PRI than that of many scholarly or popular portrayals. Though there are compelling reasons for considering the PRI and the state both coterminous and stable for much of the twentieth century, the slow but steady transition to meaningful plurality begun at least as far back as the 1980s, the many internal ruptures and antagonisms within the PRI, and compelling arguments about the nature of the PRI's always locally negotiated control in various parts of the country all point to the analytical shortcomings of positioning the PRI as an unqualified Hobbesian leviathan.[38] For Knight, the PRI's hegemony always resembled swiss cheese; it was full of holes (Knight 1990). As ensuing years would demonstrate after the transition, Mexican democracy is also riddled with issues that cannot easily be lain solely at the feet of the ruling party. Opposition parties have also offered ample evidence of their own dangerous ills, whether inherited from the PRI or cultivated afresh. The PRI, that is to say, was never quite the mythic machine its most intransigent critics envisioned, its monstrosity notwithstanding.

A third insight from Knight's response to Tuckman has been simmering beneath the surface for much of the two chapters that make up this first part of the book. In reminding readers of Mexico's "intractable social, economic and security problems," Knight broadens the focus of arguments that again place undue weight on political plurality, and indicates the position party politics find among the country's (and, even more, the city's) many pressing concerns. Explosive growth in the twentieth century brought to Mexico City a host of incredible ecological, economic, and social calamities, quite aside from those conventionally associated with electoral politics. The polymorphic monstrosity of the other leviathan of Mexico's long twentieth century, that is, merits any measure of ink spilled in homage, detraction, sorrow, and hope. Sprawling over 3,700 square kilometers, parts of el monstruo are now governed by three different states, each with its own political priorities and problems. In the summer, seasonal rains cause flooding that annually threatens lives and livelihoods, and inevitably collects in those portions of the city that tend to be both the poorest and least well serviced by the state. Ironically, the city also suffers from a lack of potable water, to such an extent that so-called water blackouts (wherein the water is shut off to portions of the city for a given period of time, usually several days) are a normal part of daily life for many. Many poor residents rely on dangerous illegal and/or unreliable connections—to aging city water pipes and to well-positioned local caciques—or simply pay what amount to exorbitant rates to have it biked or trucked to their homes in *garrafones* (water cooler jugs), probably the most common practice for rich and poor alike across the city. Climate change only exacerbates these water troubles. In the winter, thermal inversions intensify already excessive air pollution to dangerous levels, despite efforts aimed at curbing automotive and industrial emissions by several decades of PRI and PRD regimes. These efforts include industrial regulation, continued investment in public transportation (though the prevalence of decades-old *camiones*, *peseros*, and other smog-belching, individually or collectively owned people-movers competitively criss-crossing the city's constellations of neighborhoods—especially the farther one travels from the neighborhoods home to the gente decente—is suggestive of a certain degree of intransigence), and the incredibly unpopular selective restrictions on automotive traffic collectively known as *hoy no circula* (noncirculation day), which limits the days a given car can legally operate. In recent years, winter months have seen mayors double-down on this last strategy, compounding the number of cars to be kept off the streets but also posing dangerous risks associated with overcrowding to users of an already strained public transit system. Earthquakes and landslides continue to pose significant danger to many, despite the implementation of stricter building codes designed to mitigate the potential effects of such events.[39] Whether in the informal and autoconstructed peripheral zones of the city where the regulation of generations of cinderblock and corrugated metal housing was and is lax at best, or in the luxury towers of the city's premier edge city (Santa Fe)—where bribes and kickbacks are rumored to have ensured government permits even for those residential towers built atop unstable cliffs on the edges of a defunct sand mine, some of which began to signal collapse in late 2015—unstable geology continues to present an existential risk to what may even be a majority of the city's residents. The sum of these environmental risks is approached thus by Mancebo (2007, 110):

> Floods, landslides, subsidence, volcanism, earthquakes—it would be diffi-
> cult to find a more dangerous site. Mexico City is one of the most unbear-
> able and most vulnerable cities on earth, exposed to a combination of so-
> called natural hazards, poorly controlled technological hazards stemming
> from heavy industry, pollution, including an accumulation of air pollutants,
> and diminishing local resources, particularly water. These different compo-
> nents interact in series to create specific disasters. The dangers to which the
> city is exposed are euphemistically called "complex hazards." This concept
> covers a wide spectrum of risks since the urban problems resulting from the
> breakdown in the social fabric after any kind of catastrophe are part of these
> "complex hazards."

Interpersonal violence also continues to plague greater Mexico City. Though the city escaped much of the worst of Calderón's drug wars (which hit the northern states of the country particularly hard), murder, kidnap-ping, and other violent crimes continue to be a significant problem. Mex-ico's greater relative consumption of narcotics in recent years has also brought new forms of violence and despair to the city, to such a degree that as many as four thousand clandestine drug rehabilitation centers (*anexos*) now reportedly operate in Mexico City alone, many practicing a form of "violence *as* care" (Garcia 2015, 455, original emphasis) that subjects largely involuntarily committed individuals to months or years of torturous "treat-ment." Violence against women, though not perhaps matching the propor-tions of that in Ciudad Juárez along the US-Mexico border, has in recent years reached horrifying new levels. *Reuters* reported in 2014 that the mur-der rate for women in the Estado de México doubled during the second half of Peña Nieto's tenure as governor, and has been called a pandemic. Rama and Diaz (2014) describe the site of some such murders as follows:

> So many teenage girls turned up dead in a vacant field on the outskirts of
> Mexico City that people nicknamed it the "women's dumping ground." They
> began showing up in 2006, usually left among piles of garbage. Some were
> victims of domestic violence, others of drug gangs that have seized control of
> entire neighborhoods in the gritty town of Ecatepec, northeast of the capital.
> The lot has been cleared and declared an ecological reserve. But its grisly past
> is not forgotten and the killings have only accelerated.

Irrevocably linked with the party that consolidated control and directed its growth after the Revolution, the city's latticework of complex, tendril hazards and the wounds they cause together form much of the geological, ecological, social, and political substrata upon which everyday citizens attempt to construct a life in contemporary Mexico City.

CONCLUSION TO PART I

The last hundred years brought a host of changes to Mexico City. The two preceding chapters have explored this complex urban history from a geographical perspective, focusing especially on the tumultuous relationship between the country's two great monsters, the PRI and the capital city. This urban revolution—the rapid urbanization by means of which Mexico City went from the relatively small colonial capital of a highly regional and predominantly agrarian country to an immense metropolis of fabled monstrosity—also wrought inestimable new lines of conflict into the urban landscape, many of which continue to exert discernible influence on everyday residents' perceptions and actions. Though essential, these demographic and spatial expressions of the urban revolution are far from exhaustive. Mexico City's urban revolution has also been and continues to be an ecological, economic, socio-cultural, and especially political revolution, the contemporary implications of which are difficult to deny and even more difficult to see and understand if severed from the foregoing historical geography. Part II of this book is dedicated to tracing the most salient of these implications as experienced, perceived, understood, and contested by a variety of actors in the contemporary Mexican capital, ultimately arriving at their uneasy coalescence in chilangos' revolutionary structure of feeling.

Both because of how these connections are most often conveyed by residents, political operators, and others and for the sake of analytical clarity, I have chosen to conceptualize them as trajectories of conflict, by which I mean themes around and through which significant conflict has developed. Importantly, *trajectory* does not imply path dependency. Though these trajectories have determined the city's political present in important ways, neither geography nor history, is, precisely, destiny.[40] The preceding chapters have examined these trajectories in their nascency and emergence; the chapters that follow will in turn analyze contemporary expressions, relying primarily on ethnographic research.

The first of these trajectories concerns the depth but more especially breadth and variety of urban problems legible in the city beginning in the late 1960s and increasingly dire from the 1980s onward. As the foregoing chapters have begun to argue, and as the chapters that follow will more fully demonstrate, many of the dizzying array of crises that manifested as political, economic, social, environmental, etc., in nature find their origins (and many of their proposed solutions) in this urban revolution. Each and all, many analysts, theorists, and militants now argue, are most productively approached as specifically *urban* problems. Exemplary attempts to reckon with this simple but provocative proposition from the 1980s onward are the subject of Chapter 3, which follows the development of the expansive and inclusive notion of the right to the city—a political project through which to demand and build just urban futures resolutely in opposition to the logics of urbanization that fostered the above-noted crises. The second major trajectory of conflict opened up by this historical geography is more targeted, and refers to the multifaceted problems of the city's democratic processes, including especially the fractious nature of left-leaning political parties and coalitions, authoritarian tendencies and clientelist practices inherited from earlier political eras and alliances, and the overt suppression or duplicitous elision of genuine, deep, or substantive democracy by a variety of means. Chapter 4 engages with this legacy through an examination of the politics and geography surrounding an urban megaproject, the Corredor Cultural Chapultepec. Finally, the preceding material has demonstrated a consistently profound tension between scales of authority, state entities, and bodies politic, most especially those organized as the urban and the national. To be sure, such conflict existed long before the PRI, the unfathomable urban explosion of the mid-twentieth century, or the massacre at Tlatelolco. The specific shape this tension has acquired over the last century, however, weighs heavily on contemporary urban life in particular ways, exerting pressures and setting limits, as Raymond Williams might have it, on imagined political and geographical futures. Those who toil in this trajectory, as Chapter 5 explores, consistently and conspicuously couch their efforts in the language and lineage of (R)evolution.

Part II.
Revolutionary Urbanisms

CHAPTER 3

Dreaming Dialectically

The Death and Life of the Right to
the City in Mexico City

The purpose pursued in the formulation of this [Mexico City Charter for the Right
to the City] is oriented to confront the causes and manifestations of exclusion: economic, social, territorial, cultural, political and psychological. It is presented as a social
response against the city-as-commodity, and as expression of the collective interest.
Jordi Borja et al. (2006, 152)

In the end, although the catastrophe may be very real, catastrophism is the celebration of the incredulous in which irresponsibility mixes with resignation and hope, and
where—not such a secret doctrine in Mexico City—the sensations associated with
the end of the world spread: the overcrowding is hell, and the apotheosis is crowds
that consume all the air and water, and are so numerous that they seem to float on
the earth. Confidence becomes one with resignation, cynicism and patience: the
apocalyptic city is populated with radical optimists.   **Carlos Monsiváis (1997, 35)**

In the late 1980s and early 1990s, enormous political changes were afoot
in the city of palaces. The fragile coalition that nearly carried Cuauhtémoc
Cárdenas to the presidency in 1988 was splintering, and staggered to find
new footing as some factions joined Cárdenas's emergent PRD and others
established or reestablished linkages with the PRI. The PRI's patronage networks in the city were beginning to disintegrate in some areas, and were lost
to the new PRD in others. As discussed in the previous chapter, the PRI's
loss of control of the city was made possible by the convergence of several
interrelated developments. One such factor was the struggle over the PRI

itself between the rising technocratic wing (técnicos) increasingly oriented toward international economic integration and other neoliberal reforms, and the old guard políticos, many of whom still clung to the party's long-standing clientelistic practices and ISI policies. Such old guard policies had contributed heavily to a related factor of the disintegration of PRI control, the fiscal crisis of the early 1980s, fueled as it was by massive state spending, the last gasp of Mexican ISI made possible by the promise of oil revenues that never materialized in the proportions expected. Another factor was the devastating earthquake of 1985, and even more the PRI response to this tragedy. Many of the city's poorest neighborhoods were severely damaged in the quake, and the ruling party's efforts to rescue residents and make their neighborhoods livable again were heavily criticized for being slow, uneven, and politically motivated. In the lost decade, the language of crisis was stripped of much of its caustic shock, and acquired instead the trappings of a smoldering existential dread and/or rage as chilangos fought to survive the steady proliferation of highly diverse and catastrophic manifestations of the city's urban revolution seeded by previous generations of the ruling party. As frustration with the PRI grew among the city's grassroots and its disaffected residents, so too did political pluralism and resident demands for ever-greater democratization.

The disaffected neighborhood groups of the city developed in the 1980s into robust social movements, the most significant of which remains the Movimiento Urbano Popular (MUP). The MUP has its origins in the unrest of the late 1960s, but its numbers and energy swelled in the wake of the 1985 earthquake. A conglomerate entity, the MUP is made up of some twenty-eight incorporated movements whose activities span the city-region and whose leaders do not always see eye-to-eye.[1] MUP factions sometimes cooperate and participate in joint actions, but in other instances do not. Often, MUP factions join other civil society groups to jointly pursue common goals, or enroll such groups in their own actions. In the early 1990s, one such collaboration saw members of the National Democratic Convention of the Urban Popular Movement (hereafter MUP-CND) and like-minded colleagues from Habitat International Coalition-Latin America (hereafter HIC-AL) participate in a series of meetings with NGO and civil society leaders from Brazil in preparation for the 1996 UN Habitat II conference in Instanbul. It was at these meetings that the language of the *right to the city* first entered the lexicon of Mexican civil society in a major way.[2] Though both Brazilians and Mexicans would subsequently pursue an agenda they explicitly named the right to the city, they took starkly different paths. In Brazil, the right to the city took on the status, if not always the force, of constitutional law. In Mexico, the concept spread slowly through civil society networks, quietly weaving together the demands and desires of a far-flung collection of interests and steadily accruing political capital through the careful and persistent advocacy of a small group of deeply revered figures and organizations. Though it did not become law as such, the right to the city became one of the most salient and powerful organizing principles for civil society groups in Mexico City over the next several decades.[3]

In July of 2010, nearly twenty years of energy and activism around the right to the city reached a crescendo with the mayoral endorsement of the Mexico City Charter for the Right to the City (La Carta de la Ciudad de México por el Derecho a la Ciudad, hereafter Charter). The product of several years of concentrated organizing, caucusing, and debate among diverse groups and individual participants across the city, the Charter joined other similar declarations and elaborations of group demands collected under the banner of the right to the city in other cities across the globe, and currently represents the political high-water mark of the strategies expressly pursued under this name in Mexico City. Despite lacking the formality of legislation, the Charter has had a significant impact on the city's politics, and continues to be a reference point for claims-making and policy critique, and even, in a more limited way, local jurisprudence and legislative debate. Important though this moment remains in the capital's recent political history, however, the achievement of the Charter also marks the point of tidal retreat from what had been an important political strategy. In the immediately ensuing years, civil society energies shifted to strategies with other christenings as official posts were shuffled, allegiances realigned to keep in step with the unstable political substrata, and, most significantly of all, the city itself entered more deeply into a new period of its ongoing urban revolution, marked by incredible political, economic, and aesthetic alteration. As Undersecretary of Government Juan José García Ochoa explained to me in 2015, the twin dynamics of shifting priorities among government administrations and civil society groups and the changing face and economic attractiveness of the city conspired to change the equilibrium as one mayoral regime gave way to the next and the consensus built around a particular vision of urban development collapsed.

This transition away from a politics explicitly labeled "right to the city" in Mexico City coincides with an apparent academic exhaustion with the concept. This latter trend rests on a grave but common misconceptualization of the right to the city that conflates the dissipation or reconfiguration of movements with failure and reduces the radically malleable, invasive, shape-shifting insurgency of the right to the city to a flat, static, and digestible program. A deeper engagement with the inconvenient complexities of the history of the right to the city in Mexico City reveals constructions and activities that militate against such simplification, sometimes to the detriment of the objectives of the differently situated members held tenuously together in a tumultuous web of collaboration. Exploring this history is the focus of this chapter, with particular attention paid to the groups, persons, events, and processes surrounding the Charter. I will argue that to approach a full understanding of the right to the city as it has been developed Mexico City, the concept and its history must be assessed according to the dialectical principles by which it was originally conceived and subsequently reenvisioned. Properly contextualized theoretically and examined empirically, the full reach and import of the right to the city become clear, and the idea takes on a different kind of significance as the flexibility of its usage—with all its political and ideological capital *and* baggage—allows it to be understood as a vision, tool, and principle with multiple and shifting valences. In the next section, I elaborate a reading of the right to the city grounded in Lefebvre's peculiar practice of dialectics. I then use the insights developed through this reading to analyze the path of the right to the city in Mexico City from its conceptual beginnings in the late 1980s, following the politics pursued in its name through the constitutional reforms of 2016. The concluding section distils the foregoing history into two propositions: that fixing a dialectical idea is a violent act with often unappreciated consequences, and that the right to the city—by virtue of its dialectical character—need not always travel under its own name.

RECOVERING THE RIGHT TO THE CITY

Attoh's (2011) and Marcuse's (2009) recognition that instantiations of the right to the city inevitably involve compromises and trade-offs has become a basic sticking point for right to the city theory in recent years, and practice has run aground on legal or quasi-legal instruments the limits of which often leave much to be desired.[4] Worse, the concept's ease of use leaves it open to appropriations by hegemonic interests in a tragic choreography of the prophecies of generations of rights talk nay-sayers (Rorty 1996; Tushnet 1984; Žižek 2005; Brown 2004; Ignatieff 2001).[5] Perhaps owing to these developments, there seems to be a growing frustration with and desire to move beyond the right to the city in theory (Gray 2018; Merrifield 2011), and a certain sunsetting of the idea in practice.

In the gathering tumult of Paris's 1968, Lefebvre bequeathed the world an inscrutable idea bound to produce such aggravations, and imbued with the sweetest of siren songs. "The right to the city," Lefebvre (1996, 158) famously intoned, "is like a cry and a demand." It "manifests itself as a superior form of rights: right to freedom, to individualization in socialization, to habitat and to inhabit . . . to the *oeuvre*, to participation and *appropriation* (clearly distinct from the right to property)" (173–74, original emphasis). Speaking with a voice that finds its register in a language of rights then in the process of becoming globally ascendant (Moyn 2010) and simultaneously in an unmelodic anti-harmony that viciously mocks and undermines at every turn this bourgeois program of hardened legal norms as the desiccated, hollow promises of the state beholden to capital and the expansion of the spatial sway of its crucial abstractions (e.g., "man" and "citizen"), Lefebvre's mystifying rejection/assertion of rights beckons all comers. In the face of and in brazen contradistinction to the grandiloquent declarations of states and sovereigns, *The Right to the City* and its central concept were touchstones meant to serve as a cross-cutting incision into a world of analysis in which the urban did not sufficiently figure, as grist for later work on capitalist urbanization, and as breadcrumbs for those who would pursue this struggle, or, as Buckley and Strauss (2016, 632) describe Lefebvre's works on the urban, "invitations to revolutionize the day-to-day practice of intellectual knowledge production about it." At once a frank dismissal of rights and a bold claim for them, the right to the city abides in contradiction. Its full expression is most visible through an approach capable of appreciating the complex web of material and virtual connections it seeks to inhabit and its perpetual internal motion driven by the forces that stretch, twist, sever, and bind these ever-proliferating ties. His right to the city is multivalent, slicing through space and time synchro-spatially in its analysis of urbanization and the gestures and demands it throws out toward an unalienated urban future. Viewed as a solid, finished abstraction as presented by Lefebvre and others to follow, the concept presents no particular problems for even a casual observer, and, as recent academic and popular trends have demonstrated, continues to draw a large audience. Asked to function in undialectical fashions, however, the right to the city is often found lacking, its seemingly endless efforts to evade fixity frustrating even the most sincere efforts to imbue the concept with the force and duty of law.

Placing the right to the city in a Marxist dialectical lineage is hardly to assign it a predictable form or function, however, given the divisive role dialectical reasoning continues to play in Marxist and non-Marxist thought (Castree 1996; Dixon, Woodward, and Jones, 2008; Jameson 2009; Ollman 1971, 1993, 2003). Arguing against common misconceptions, Ollman (2003, 15) positions dialectics thus:

> Dialectics is not a rock-ribbed triad of thesis-antithesis-synthesis that serves as an all-purpose explanation; nor does it provide a formula that enables us to prove or predict anything; nor is it the motor force of history. The dialectic, as such, explains nothing, proves nothing, predicts nothing, and causes nothing to happen. Rather, dialectics is a way of thinking that brings into focus the full range of changes and interactions that occur in the world.

Changes and interactions are both key to Ollman's explication of dialectics à la Marx, and here name two major pillars of dialectical investigations in this variation. For Ollman, whose corpus on dialectics spans several decades and provides the most thoroughgoing and sustained methodological engagement with the subject, dialectics is first and foremost a science of movement and change.[6] This approach treats mutability, contingency, and inconstancy as axiomatic, and thus constitutes what Dixon, Woodward, and Jones (2008, 2549) call a process philosophy, a consideration of a world perpetually in a state of "continuous becoming." Ollman (2003, 66) roots this epistemological priority of movement over stability in Marx, "so that stability—whenever it is found—is viewed as temporary and/or only apparent, or, as he says on one occasion, as a 'paralysis' of movement." Established on a fundamental rejection of ontological stasis in the Aristotelean logical lineage, dialectics addresses itself instead to morphosis. Engels (1940), in a conceptualization of dialectical materialism Jameson (2009, 13) calls "far from outmoded," identified three laws of dialectical movement: 1) "the law of the transformation of quantity into quality and vice versa"; 2) "the law of the interpenetration of opposites"; and 3) "the law of the negation of the negation." To these, which he affirms at length, Ollman (1993, 2003) adds several other types of dialectical movement, or ways of understanding such movement, including metamorphosis, contradiction, mediation, precondition and result, and unity and separation. Dialectical movement is largely considered synchronic and nonlinear (Castree 1996) and propelled by "contradiction, reinforcement, fragmentation and reconstitution" (McFarlane and Silver 2017, 459; see also Harvey 2014). Apprehending in ways that transcend linear time requires the aid of a memory like that possessed by the Queen of Hearts in Lewis Carroll's *Through the Looking Glass*, a memory that "works both ways."[7] Dialectics from this vantage can thus be likened to the fictional language Heptapod depicted in Denis Villenueve's 2016 film *Arrival*, the translation of which allows the protagonist to access a global set of past and future human knowledges perhaps best approximated by psychoanalyst Carl Jung's controversial collective unconscious. Such a synchronic perspective is essential not only to Marx's analysis of the mode of production, for example, but also to his mode of presentation, wherein the reader must be able to follow conceptual developments that unfold in more than one direction through a text and change through interaction with an expanding and often shifting conceptual universe (Castree 1996; Henderson 2013).[8]

Even the most creative of dialectical investigations, however, often fail to fully realize the methodological utility of dialectics for addressing the future. Though Ollman (2003, 17) insists that "the future finds its way into this focus as the likely and possible outcomes of the interaction of . . . opposing tendencies in the present," it is the likely rather than the possible outcomes of such processes that receive the bulk of the attention in their work. A common and understandable tendency, such preference runs the risk of casting anemic abstractions by virtue of static projections of the future, resigning the future itself to the approaches of Carroll's Queen of Hearts or *Arrival* protagonist Louise Banks, for both of whom it is a domain of only *memories*.[9] This tendency, too, is rooted in a reading of Marx (1970, 21), who infamously states, "Mankind thus inevitably sets itself only such tasks as it is able to solve, since closer examination will always show that the problem itself arises only when the material conditions for its solution are already present or at least in the course of formation." Read simply, this statement captures dialectical synchronicity in its most mechanistic, nomothetic expression, wherein forecasting focuses on the probable solutions to only those problems already at least partially extant. As McCormack (2012, 729) argues, however, "to experiment with abstraction is not to move thinking away from the material: it is to do something with material effects." Abstraction, as a process that "has as its object the unknown unknowns of potential futures" (McCormack 2012: 728), can and does play a significant role in shaping present and future realities by not only illustrating but also, more importantly, by *creating* virtual realities. In Marx's (1967a, 174) likewise infamous statement on the fundamental distinction between the "worst architect" and the "best of bees," such creative abstraction is highly prized.[10] The dialectical imagination is thus a powerful tool not only for speculation on the future but also for the production of the future (Patomäki 2017), as a generative agent constantly populating the realm of the possible and urging select possibilities toward the preferential status of the probable.

This is the inconstant morphology upon which Lefebvre constructed his *Right to the City*. Shifting modes within the text—sometimes in ways that readers continue to find frustrating—he moves from incisive critique of past and present to virtual abstraction of negated-negation, from unalienated urban oeuvres to reordered centralities of unsegregated encounter. The Lefebvrian dialectical varation is thus unusually creative, deviating from the Marxian mean in its assertion of a third dialectical energy, a creative actant forever disrupting bipolarities and forging new paths in virtual and material space. It is tempting to read in Lefebvre's triadic movement a rearticulation of Marx's problematic, as he himself sometimes explained his approach.[11] As Kofman and Lebas (1996, 9–10) argue, however, "Lefebvre's dialectic is not that of Hegel, thesis-antithesis-synthesis, nor one of affirmation-negation-negation found in Marx, but a much more open, ended [sic] movement, bringing together the conflictual and contradictory, and linking theory and practice," which "dissolves stable morphologies to such an extent that stability becomes a problem." A dialectics of praxis and becoming (Lefebvre 2003b), Lefebvre's right to the city is a special kind of abstraction. It appears as a materialization or moment of the third energy of Lefebvre's dialectical triad, his own three-headed methodological hydra. It is in one sense an abstraction cast from the patterns and processes of capitalist urbanization he investigated, a valence in which it operates as both explanation and critique. In another valence, however, it functions in "abstract virtuality" (McCormack 2012), a mode in which it can be used not only to develop and test new potentials and powers, but also to work toward their realization. The phrase morphs and slides between these modes, blending the work of critique, explication, creation, and practice, and refusing any clean separation between them. This is how and why the right to the city is both a cry and a demand, both screamed into the virtual void and fought for or practiced in the realm of the possible/present. Examining some salient features of Lefebvre's right to the city as it moves through these modes of operation will help put contemporary instantiations of the concept in proper perspective.

A useful entry point into Lefebvre's right to the city is the important—if familiar—provocation that Lefebvre was explicitly not interested in the city. Rather, the target and object of his theorizing was the urban. The city he knew best and which served as the basis of his investigations was a product of twentieth-century capitalist urbanization, and developed according to the contradictory impulses of the profit imperative by way of the exploitation of labor expanded to mass-scale. It was Paris of the 1960s, a city with an historic center increasingly scrubbed of its underclasses and an emerging *banlieue* to which they were actively being expelled. The exploding cities of Latin America, Mexico City perhaps most of all, were beginning to exhibit similar spatial arrangements according to similar logics, though by very different means (Beyer 1967; Connelly 2003). This geography, for Lefebvre, bore out the domination of urban use values by exchange value, in a familiar logic of spatial exploitation. Moreover, his exploration of the dynamics of industrial capital's ravaging and renovation of traditional urban cores and the associated spatial effects—twin processes described variously as "explosion-implosion," "condensation-dispersion," or "de-urbanized urbanization" (Lefebvre 1996, 77, 123)—prefigured what he would two years later term the urban revolution, whose dark coronation of urbanization as the primary driver of accumulation over and above industrialization would so provoke a young David Harvey, then himself in the throes of his Damascus Road to Marxism (see Harvey 1973). This industrialization-urbanization tore through old spatial centralities to create new arrangements more suited to a changing political economy, such that while old central city districts were becoming caricatured sites for the collection of consumption revenues, the expulsion of the poor and working classes acts as a safety valve in the process of the exploitation of labor by removing to a safe distance the potential for organizing and activism among these problematic elements. Thus spatially segregated, Lefebvre's city was also built upon a multivalent alienation. Its uses largely ordered and prescribed and its users and makers categorized and collated, the modern capitalist city was trending away from its potential as a space of generative encounter with difference. This is the city that Lefebvre's right to the city explicitly rejects, the capitalist city, the result of a bloodless urbanization for and by expanded and expanding accumulation. But more than this, Lefebvre also rejects the old Paris, the specter of which even then had begun to haunt the prettified districts as "only an object of cultural consumption for tourists, for aestheticism, avid for spectacles and the picturesque" (Lefebvre 1996, 148). Though the particular contents of this trend find expression from within the "structures of feeling" (Williams 1977) pertaining in distinct geographies, whether Mexican and Chicano art in Chicago's Pilsen (Wilson, Wouters, and Grammenos 2004), the aesthetic cult of exposed brick in Brooklyn (see Clark 2015), or the overtly seductive hints at the magical properties of Indigenous ingredients worked into artful, delicious, and expensive cocktails in Mexico City (see Ayrelan Iuvino 2017), it seems not only to have held but indeed accelerated across the gentrifying urban world. As a critique of such investment patterns and associated (cultural) appropriations, Lefebvre's right to the city "cannot be conceived of as a simple visiting right or as a return to traditional cities. It can only be formulated as a transformed and renewed *right to urban life*" (Lefebvre 1996, 158, original emphasis).

Still, the urban remains an elusive target for Lefebvre, as for those in his wake. Part of this difficulty is rooted in the duality of this particular abstraction. For Lefebvre, the urban refers to both the processes of capitalist urbanization and also the latent possibilities of this totality. That is, the urban here is cast with both actual and virtual contents. The majority of *The Right to the City* is devoted to an examination of the former, while the tantalizingly suggestive quotes isolated by most contemporary academics interested in this concept illustrate those few moments in the text when Lefebvre allows the future to push through the tired seams of the present. Mitchell (2011) illustrates this tension well by way of the so-called tent cities of the United States, in claiming that "what must be fought for, in other words, is not only the production of tent cities, but especially the destruction of a system that has made them an inevitable part of the urban landscape," a struggle that would at once insist not only on their right to exist but also and more importantly on a different kind of urbanization, on processes of urban development that would "make them superfluous, rather than necessary." In the first instance, then, Lefebvre is after that sense of the urban in which cities are made according to the dictates of exchange value and the accumulation of capital, along the lines proposed by Harvey (2008, 37), who proposes the right to the city as "greater democratic control over the production and utilization of the surplus." To exercise the right to the urban from this angle would necessitate a fundamental reorganization of political economy beyond a given locality, though this would seem an obvious place to start.

In a second sense, the urban refers to a largely undefined field of virtual possibility often specified only in the negative, or as the negation of the urban in the former sense. The two are thus internally related, inextricably linked, and co-constitutive, as the exercise of the right to the urban in the second, virtual, and future sense is impossible without its first, material, and present expression. This is precisely why Lefebvre's rejection of the city was only partial, despite his seeming insistence to the contrary. His call for "a renewed right to urban life" over and against the city is a rejection only of the city as then constituted, and a demand for a city that can be radically other. To borrow from his own context, Paris was for Lefebvre a material city worth desiring based on its potential to be radically remade according to practices not subject to the logic of capital accumulation. In this second register, the right to the city entails the collective ability of the city's producers and residents to craft the city, as Harvey (2012, xvi) puts it, "after their heart's desire." Lefebvre's few statements about this virtual terrain suggest that this urban future would retain many of the material qualities of the modernist city, albeit in unalienated manifestations. Notable among these are notions of mediation, difference, and centrality, which run through Lefebvre's analysis in both their alienated material forms and, occasionally, in their virtual guises.[12]

A first cut at urban mediation brings into focus the privileged position of cities as spaces of concentration for increasingly global flows of information, money, goods, and people (Sasson 2001, 2012). As Blokland et al. (2015, 655) concisely explain, the urban is "a specific sociopolitical and institutional setting, in which various scales—from the local to the transnational—are layered, condensed, and materialized." That is, contemporary cities draw to themselves immense material and virtual resources, which in turn pull on webs of connections to the realms of production, exchange, and consumption, themselves subject to any number of geopolitical, financial, and other entanglements. Cities thus become the point of interface among these worlds, the places where the virtual becomes real and the real becomes virtual, where the global and the local converge and collide, and where time and space are expanded and collapsed at ever-increasing rates as value chases after its momentary realization or ultimate destruction.[13] This positions the city, first, as mediator between scales of economic authority and processes of production. Lefebvre carried this point further, arguing that this quantitative phenomenon had achieved such a degree of significance as to constitute a qualitative change. He argued that the urban now figures, per se, as a factor in the global production process: "by grouping centres of decision-making, the modern city intensifies by organizing the *exploitation* of the whole society (not only the working classes, but also other non-dominant social classes). This is not the passive place of production or the concentration of capitals, but that of the *urban* intervening as such in production (in the *means* of production)" (1996, 109–10, original emphasis). Indeed, beyond the command functions of global financial capital arguably long-concentrated in cities, urban space itself has in recent years become a preferred investment target and safe haven for capital of various origins, including so-called dark money (see Story and Saul 2015; Wainwright 2014). Concomitant reorganization of state processes and spaces constitutes an important correlate to this mediation of economic and social activity. As Brenner (2000, 366) argues, "the urban scale is not only a localized arena for global capital accumulation, but a strategic regulatory coordinate in which a multiscalar reterritorialization of state institutions is currently unfolding." Not least in Latin America, the realization of such reterritorialization of urban state space has encouraged academics, activists, and policymakers to use the language of the right to the city as a way to productively engage the metropolitan question, especially as sprawling megacities outgrow their political boundaries in waves of expansive settlement.[14]

Their concentrations also enable the production of centrality in cities, another multivalent term in Lefebvrian analysis. Aside from these already noted agglomeration effects long under the purview of the most disinterested scientific approaches of economic and urban geography, Lefebvrian centralities entail the full expression of urban use values. Notable among these is Lefebvre's notion of the oeuvre, an amorphous concept whose literal translation is "work." From this vantage, both the urban itself, writ large, and the city in particular, are not only products but works, as in the sense of works of art (Mitchell 2003; Purcell 2013). Such works are the collective expressions of urban residents and urban workers, composed of everything from the architectural and artistic feats of its highest-order artisans to the quotidian secretions and enactments of ordinary culture and the all-important rhythms of everyday life.[15] Foregrounding the role of common activities in the production of the oeuvre, Lefebvre's notion of centrality is shaped by an approach to the city quite different from that of other prominent Marxist urbanists, such as Castells (1983), whose interest centered on collective consumption, or Harvey (2006), whose focus was the role of urban space in production crises (Kipfer, Saberi, and Wiedetz 2013). It is in this way that the right to the city becomes, according to Purcell (2013, 561), "a conjoint claim by the users of urban space to take greater control over its production." Just any role in this production of the oeuvre will not do, however. Lefebvre's insistence on centrality, heavily influenced by the expulsion of the working classes from central Paris, entails a claim both to significant influence in the production of urban life and to the physical spaces of cities. As Mitchell (2003) argues, access to the latter conditions access to the former of these claims. Taking exemplary material from the urban United States, especially with regard to the plight of homeless populations, Mitchell convincingly demonstrates that the bundle of citizenship rights often taken to be part and parcel of the right to the city mean little if their bearers can simply be expelled from the city legally beholden to them.

In ascribing such importance to centrality, Lefebvre's right to the city also affirms the role cities have long played as the stage of encounters with difference. In this instance, the right to the city operates not only as an investigative and explanatory vehicle for demonstrating how cities have historically facilitated such encounters, however, but also as an explicit critique of the erasure of difference in the city of his day. As he would later argue more fully and forcefully in *The Production of Space*, this critique is addressed in part to the proliferation of abstract space. "Abstract space," Lefebvre (1991, 49) explains, "functions 'objectally,' as a set of things/signs and their formal relationships. . . . Formal and quantitative, it erases distinctions, as much those which derive from nature and (historical) time as those which originate in the body (age, sex, ethnicity)." This is the space of defined uses and settled meanings, where consensus all but guarantees adherence to a set of rules that govern spatial practice. Though his abstract space was that of capitalism, of urban planning, and of the capitalist state, it nevertheless remained for him a complex category that "has nothing simple about it: it is not transparent and cannot be reduced either to a logic or to a strategy" (Lefebvre 1991, 49). Perhaps its most salient feature is its dialectical tension with the differential space it is forever attempting to erase or suppress. In its proliferation, abstract space papers over different interpretations, meanings, and uses of space in much the same way that Lefebvre's urban fabric is said to spread urban relations to an ever-greater share of the earth's surface (Lefebvre 1991, 1996, 2003a), and, once instantiated, attempts to suppress moments of differential space, "which it carries within itself and which [seek] to emerge from it" (Lefebvre 1991, 50). Conflicts of abstract and differential space find countless material expressions, from anti-homeless measures in US cities (Mitchell 2003) and the clearing of the Occupy protesters from New York's Zucatti Park (Moynihan 2012), to the disappearance of traditional informal services performed by neighborhood figures in increasingly desirable areas of Mexico City (Morse 2015). In its virtual guise, the right to the city is a valorization of differential space and the potential its expressions may hold. Elsewhere articulated as a right to difference, Lefebvre is here interested in the possibility of a variety of expressions of difference, from those corporally embodied to those realized in alternative practices in and of space, "of the prioritization of the lived over the conceived" (Merrifield 2006, 115), and of "representational spaces" over and above "representations of space" (Lefebvre 1991).[16]

This irrepressible right to difference has had a notable impact on the organization and orientation of struggles for and articulations of the right to the city in practice. Most importantly, the valorization of heterogeneity has encouraged the formation of coalitions of interests that often vary considerably. The Brooklyn-based Right to the City Alliance, for example, boasts some forty-nine-member organizations and twenty-three allies across the United States, whose nominal purviews include such wide-ranging issues as racial and ethnic justice, tenants' rights and anti-eviction organizing, environmental justice, education, and neighborhood and community empowerment (Fisher et al. 2013; The Right to the City Alliance 2016). Marcuse (2009, 191) lauds the special potential of the right to the city to bring together interests often considered divergent in the extreme, even suggesting that "it is a combination of the deprived and the discontented who will lead the push for the right to the city."[17] In advocating for such broad coalitions as a way to propel the right to the city forward in disparate contexts, Brown and Kristiansen (2009, 37) emphasize the need to collect, unite, and strengthen existing social and political resources and groups, as they put it, "drawing together existing strands." Purcell (2013, 566) likewise sees such potential in the right to the city, but as a "conjoint claim" rather than a collective one, a conceptualization that "emphasize[s] that the mobilized entity, the claimant of rights, consists of multiple bodies joined together (conjoint), rather than as a single, unified body (collective)." For Purcell (2013, 572), such claimants can and should establish "networks of equivalence" in pursuit of a right to the city conceived of as a process rather than an end. Purcell's equivalence (following Laclau and Mouffe 2001) would thus allow movements to steer toward relations with difference resembling those of a dialectical right to the city, in which the theoretical notion of identity/difference is realized through a union of interests that refuses both interest-driven hierarchies and the sublimation of identities.

Lefebvre's triadic musings on the crucial site for this right—the urban as the level of mediation, as the seat and site of centrality, and as "the place where difference lives" (Mitchell 2003, 18)—owe much to his experience of both Paris and Navarrenx, the medieval town of his birth on the northwestern edge of the French Pyrenees (Merrifield 2006). While the Parisian expulsion of the underclasses propelled his insistence on the right to centrality and the primacy of what the great modelers of the second city were already calling the CBD (Central Business District) as the center of power and the privileged place of encounter, haunting memories of the provincial harvest festival surreptitiously waged an insurgent campaign of free play, experimentation, and unproductive consumption.[18] Meandering along both currents, Lefebvre's right to the city is able to make demands of a legal and political order (a capitalist, liberal-democratic state) it simultaneously decries as destructive to the human species and its environs, sowing and tending the landscape for moments of urban emancipation from this very political economy. In practice, this has led to the aforementioned difficulties in defining, instrumentalizing, and even assessing the utility of the concept. There are those who would see it more carefully specified in law (Fernandes 2007; Parnell and Pieterse 2010), those for whom its "capaciousness" or "strategic fuzziness" remains a strength (Attoh 2011; Mitchell and Heynan 2009), those who consider its potential compromised (Merrifield 2011; Uitermark et al. 2012; Vradis 2012), and those who continue to see in it a potentially transformative politics (Fisher et al. 2013; Harvey 2008; Leontidou 2010; Marcuse 2009; Purcell 2013, 2014). A dialectical vantage reveals both the legitimacy and the limitations of all such perspectives. Considering the right to the city as an analytical tool, debating its political utility, and seeking to implement it in practice will inevitably desiccate or imprison an inherently fluid abstraction, freezing the motion of a dialectical process into what can most aptly be described as a dialectical image (Wright 1999). Attoh's (2011) well-argued claim that scholarship on the right to the city ought to begin with the recognition of Waldron's (cited in Attoh 2011, 679) proclamation that "the institutionalization of any right [particularly within a world characterized by scarcity and conflict] poses tradeoffs" thus misses the mark by half, as does Mitchell's (2003) emphasis on Marx's formative attitude toward rights.[19] While the necessity of difficult choices is a point well taken, this obligation is rooted not only in the nature of the right to the city as a right inevitably in conflict with other rights and shackled to the structures of political economy, but also fundamentally in the nature of any dialectical abstraction.

Building on these insights, my analysis departs from the common and mistaken analytical thrust to figure out or uncover the truest hard kernel of the right to the city.[20] Beginning from principles of plurality, contingency, and inconstancy, in the following sections I examine the complicated political life of the right to the city in Mexico City, including its appearances, disappearances, and reappearances in recent decades. Understanding why and how the right to the city rose to political prominence in Mexico City, its shifting valences and purported political demise, and the significance of its recent reemergence requires an analysis attentive to its dialectical nature and heritage, an analysis capable of appreciating its morphology and movement through regimes, coalitions, moments, forms, and functions. As the following sections will show, proponents of the right to the city in Mexico City have made admirable attempts to craft a radically malleable vision through the modifiers *complex* and *collective*, but even this careful praxis has run afoul of the violences of abstraction as the rapacious forces of (re)development remake the city to which they address their critiques and demands. Understanding the right to the city dialectically, however, will also illustrate the mutability of even the most dire states of stasis, putting into proper perspective both the consequences of these urban politics and the revolutionary potential the idea still holds in Mexico City and beyond.

REALIZING THE RIGHT TO MEXICO CITY

MUP and HIC-AL leaders of the late 1980s and early 1990s found in the right to the city a concept more applicable than any other to problems then confronting Mexico City.[21] The idea presented them the opportunity to elaborate connections they had long seen between burgeoning crises conventionally considered in isolation within the atomistic ambits of the Federal District's already enormous bureaucracy. Like the MUP itself, the right to the city promised to integrate disparate issues and voices, to illustrate the links between, to choose just one example, two decades of turbulence in the regional political economy and the ravages wrought on the city's air quality, transit system, and water and waste infrastructures. As HIC-AL ex-president Enrique Ortiz Flores (2008, 1) would later explain:

> The right to the city is located at the center of great contradictions and the highly dynamic interactions generated in contemporary society. From the positioning assumed and the effectiveness of the paths followed to address these phenomena will emerge the possible organization of the city as space of collective co-existence and of viable and sustainable good life for all of its inhabitants.

In a similar vein, future HIC-AL President Lorena Zárate (2011, 269) would also later add:

> As a complex right focused on a highly populated territory of multiple relevance for the country and with severe pressures on environmental conditions, the right to the city must propose a vision that surpasses the specialized approaches of distinct disciplines, professional practices, and the structure of public administration, as well as the individualistic and consumerist attitude prevailing among a large proportion of inhabitants.

Its capaciousness, however, was far from the only thing that attracted early proponents to this language. As it was taken up in Mexico City, the right to the city created harmonic resonance with other demands and struggles then finding their form in the language of rights, in a political climate that was in the early stages of radical realignment. Indeed, as Zárate (2011, 269) continues, "Human rights and democracy are not abstract phenomena; they are attributions and processes of specific people in specific places. As we conceive it, the right to the city can and should also be a tool through which to territorialize the former and deepen the latter." As calls for democratization gained momentum, including especially pleas for the political autonomy for the capital city, for ridding the political process of fraud and corruption, and for ending the reign of the fracturing PRI, rights talk gained considerable political purchase.

Though the common story in Mexico City civil society has the right to the city emerging from conversations, meetings, and forums convened between the 1992 Rio de Janeiro Earth Summit (The United Nations Conference on Environment and Development) and the 1996 Habitat II Conference (the Second United Nations Conference on Human Settlement, or City Summit) in Istanbul, the concept first began to emerge at HIC-AL and related local organizations as early as 1989 (Ortiz Flores 1990), especially in the writings and activities of Enrique Ortiz Flores.[22] An architect by training and lifelong participant in and leader of Mexico City civil society organizations, Ortiz has amassed considerable credibility and political capital among diverse groups ranging from residents of the city's peripheries to the mayor's office and the Association of Engineers and Architects of Mexico (AIAM).[23] When I met him in 2014, Ortiz was serving as past president of HIC-AL, based in Mexico City's slowly gentrifying Roma Sur, just off of the crowded north-south arterial known as the Avenida de los Insurgentes (Avenue of the Insurgents), reportedly one of the world's longest avenues and home to the flagship line of the city's Metrobus system. Though already nearly eighty years old at that time, Ortiz still maintained a schedule packed with government and civil society meetings, interviews with press outlets, research activities on projects that span the gamut of local to global urban issues, phone calls, and writing. A prolific author, the research reports, position papers, editorials, essays, and other manuscripts he has penned over his many active decades can fill a small library (and do, as I can personally attest), and he works comfortably in English and Spanish. He has for decades been motivated by ideas of autoconstruction at the city's peripheries and the social production of habitat, and by an expansive notion of rights. On the one hand, he has long advocated for the rights guaranteed to the city's residents by the various levels of the Mexican state. On the other hand, he has also pursued rights that are far more complex than any straightforward juridical sense the term *right* may convey.[24] His advocacy of the right to the city spans and draws upon his time spent as the head or leading member of HIC-AL and a number of other local organizations, his deep personal and working relationships with a wide web of civil society groups (especially the leadership of some segments of the MUP), his collaborations with the National Autonomous University of Mexico (UNAM), the Autonomous University of Mexico (UAM), and other national and international universities, and his longstanding connections to various offices, personages, and arms of the local and federal state.[25] In meetings with local civil society or government functionaries, with transnational funding agencies or international nongovernmental organizations,

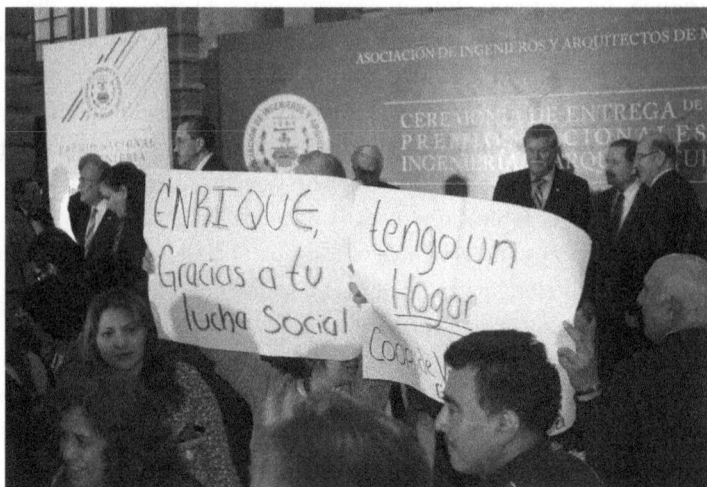

FIGURE 3.1. Signs reading "Enrique, thanks to your social struggle, I have a home" displayed at an event honoring Enrique Ortiz Flores, who was awarded the National Prize for Architecture. Photography by author.

Ortiz's analysis and opinions carry significant weight, and his name alone can open a great many doors in Mexico City.

An early proponent of the right to the city, Ortiz has long considered it a right both complex and collective, asserting at once a universality of application and belonging and a composition that requires negotiation in practice (Ortiz Flores 2008). For Ortiz, the right to the city applies and properly belongs to all residents of the city, and moreover to all human beings by virtue of its consideration as a human right. Its complexity is related, in his formulations, both to its malleability and limitless connections and relations to and with other rights and issues, and to its status as a right which contains, covers, ensures, and/or protects other rights and relations (Ortiz Flores 2008). This complex nature goes a long way toward explaining the deep entanglement of the right to the city and other rights and principles to be found in his work, especially in the late 1980s and early 1990s, such as "el derecho a la vivienda" (the right to housing) and "el derecho al habitat" (the right to habitat) along with urban dwellers' rights to "improve their quality of life" and "participate in the planning and development of their habitat." Moreover, contemporary instantiations of the right to the city such as that found in the Charter bear striking resemblance to his early formulations, whether called by the name of the right to the city or not. For example, the six strategic areas of the Charter (discussed below) thematically align fairly tightly with seven rights Ortiz outlined in a report concluding a June

1995 meeting of "more than 600 civil organizations," including parts of the MUP, aimed at producing a "charter of citizen rights" (Ortiz Flores 1995, 42), and later a "Charter of Rights to the City and to Housing" (Carta por los Derechos a la Ciudad y la Vivienda). Ortiz's influence over the conceptual development and the political deployment of the right to the city in Mexico City can thus hardly be overstated. He has advocated for the concept's inclusion in policy conversations and popular movements for nearly three decades, and is one of the primary vectors through which countless persons in the city have come to understand this notion. His political influence also ensures that the right to the city retains a perhaps unexpected relevance in contemporary politics.

The bulk of the credit for advancing the right to the city politically, however, is usually given to the MUP, and to Jaime Rello Gómez, leader of the Unión Popular Revolucionaria Emiliano Zapata (Revolutionary Popular Union of Emiliano Zapata, hereafter UPREZ) and the Congreso Nacional Democrático (National Democratic Congress of the Urban Popular Movement, hereafter MUP-CND) in particular. A priest by training from a middle-class background, Rello has lived and worked in some the city's poorest areas, such as the infamous Nezahualcóyotl in the neighboring Estado de México. As with Ortiz, Rello's reputation is weighty, and the respect he and his opinions are shown have undoubtedly played a role in garnering support for the right to the city for several decades. In meetings with high-ranking government ministers, civil society leaders, academics, or angry residents, rooms go quiet and attention is paid when Rello speaks. Moreover, the political capital exercised by the MUP was important for bringing local government to the table in 2007 and 2008, and the MUP has been credited with much of the organizing work that lead to the Charter's development (Adler 2015a, 2015b; Wigle and Zárate 2010). Indeed, it is through the MUP that the right to the city took on force as a political entity in Mexico City. Speaking at the July 2010 public ceremony for the Charter, Rello (2010) called its signing, "possibly the clearest instrument for continuing a long-awaited dream: converting this piece of land, with so much history, into an entity, a state with its own constitution. Our great city must receive that which corresponds to it. A city of rights for all. There is no going back." In this telling statement, Rello articulated two agendas for the right to the city.[26] The second is the familiar goal of a city of rights for all found throughout the academic and popular literature on the right to the city, and in the body of work produced in Mexico City in conjunction with HIC-AL and Enrique Ortiz Flores. The first pursuit, however, turned out to be the operative one in the context of the Charter's development, by linking the right to the city to demands for political reform that were reaching a fever pitch near the end of the new millennium's first decade.

The lynchpin for transforming the right to the city from simply a favorite concept of notable activists into a robust political agenda and instrument was the city's fifth consecutive PRD jefe de gobierno, Marcelo Ebrard Casaubón. In interviews and casual conversations, I was repeatedly (and nearly unanimously) told that Ebrard had strong working relationships with Mexico City's civil society organizations. His willingness to attend meetings in the offices of NGOs (such as HIC-AL) rather than insisting that any and all meetings be held behind closed doors in the vast chambers of the state was often cited as exemplary of his attitude toward cooperation and his openness to ideas and proposals originating outside his party or administration. Collegial cooperation and a genuine spirit of progressivism, however, were not the primary reasons (despite some claims to the contrary) that Ebrard became interested in the right to the city. As he explicitly stated at several public events (Ebrard Casaubón 2008, 2010), his interest in the right to the city was primarily as a vehicle for realizing the Federal District's political reform, as discussed in Part I. Ten years after the city elected Cuauhtémoc Cárdenas its first jefe, Ebrard and other powerful voices demanded even more political autonomy for the city, including constitutional recognition as an autonomous, self-governing state within the Mexican Union (and thus an end to federal management), along with a constitution. In talks with MUP-CND leaders and others in 2007, Ebrard began to consider the right to the city a viable means by which to pursue precisely these ends.

In January of 2008, the World Social Forum convened in Mexico City's Zocalo, supported by funding from Ebrard's government.[27] One of the tents at this event was devoted to the idea of the right to the city and to habitat (Zárate 2011). Influenced by these activities, Ebrard agreed that a Mexico City Charter for the Right to the City was warranted.[28] In participation with MUP-CND leaders, the Ebrard government formalized the Comité Promotor de la Carta de la Ciudad de México por el Derecho a la Ciudad (Promotional Committee of the Mexico City Charter for the Right to the City, hereafter CPCCMDC) in April of 2008, including both Jaime Rello Gómez and Enrique Ortiz Flores. Along with the MUP-CND (who is officially credited with its integration), the CPCCMDC also included the Federal District's Secretary of Government, HIC-AL, the Federal District's Human Rights Commission (Comisión de Derechos Humanos del Distrito Federal, hereafter CDHDF), the Coalition of Civil Society Organizations for Economic, Social, and Cultural Rights (Espacio de Coordinación de

Organizaciones Civiles sobre Derechos Económicos, Sociales y Culturales, or Espacio DESC), and the office of the Attorney General for Social Affairs (la Procuraduría Social del Distrito Federal). According to Lorena Zárate (2011, 264), then Regional Coordinator of HIC-AL, the CPCCMDC conducted over thirty-five meetings throughout 2008–2009, "to coordinate, discuss, systematize, and draft the contents of the charter and to monitor and evaluate the process." To promote the process and circulate ideas, the CPCCMDC created "pamphlets, a blog, leaflets, and a video specifically oriented to animate the process," and committee members also offered interviews to media outlets and participated in roundtables, workshops, and conferences (Zárate 2011, 264). By January of 2010, CPCCMDC members reported that over three thousand people had participated in some part of the Charter's elaboration. These included laypersons solicited for interviews at public events, and academics, activists, and government officials who had spoken in meetings or offered written proposals, including at local events attended by Ebrard. The CPCCMDC also drew upon documents and ideas previously developed by other bodies, such as the inaugural World Assembly of Inhabitants held in Mexico City in 2000 (Zárate 2011), the Global Charter for the Right to the City first elaborated at the inaugural World Social Forum in 2000 in Rio de Janeiro, the 2008 Program of City Education and Knowledge, the 2009 Program of Human Rights of the Federal District, and the 2008 Diagnostic, a report on the state of human rights in the city produced by the CDHDF (CPCCMDC 2011).

The Charter, signed by Ebrard and other ministers in a public ceremony at the Municipal Theatre on July 13 of 2010, spans some fifty-seven pages exclusive of its associated materials and list of signatories, and is divided into four sections: a preamble which outlines the Charter's objectives and places the document in historical and political context; chapter 1, which defines the right to the city and the entities involved in its pursuance in Mexico City; chapter 2, which lays out the six strategic foundations of the Charter and nine guidelines for implementation; and chapter 3, which, among other things, posits standards by which violations and successful implementation of the right to the city may be judged.[29] The bulk of the text is given over to the explication of the six foundations of the right to the city as elaborated in chapter 2 (CPCCMDC 2011; Wigle and Zárate 2010):

1. Full exercise of human rights in the city; for a city of human rights
2. The social function of the city, of land, and of property; for a city for all: inclusive, solidary, equitable
3. Democratic management of the city; for a politically active [participatory]

and socially responsible city

4. Democratic production of the city and in the city; for a socially productive city

5. Sustainable and responsible management of environmental, cultural and energy resources as common goods in the city; for a viable and environmentally sustainable city

6. Democratic and equitable enjoyment of the city; for an open, free, critical, and enjoyable city

To be sure, the breadth of issues covered by these six areas of focus is reflective both of the particular makeup of the CPCCMDC and the collection of contributions it solicited from the public in the course of drafting the Charter and of the growing set of environmental, political, and other problems then confronting Mexico City. The fundamental role played by human rights in the document, for instance, can be traced to the important institutional participation of the CDHDF in the CPCCMDC, to the earliest writings of Enrique Ortiz Flores, and to the attention increasingly paid to human rights as the violence of President Felipe Calderón's sexenio spread from the rural domains of the large drug cartels to the country's major cities and the tally of missing persons soared to over twenty thousand, and those violently killed during his term to over one hundred thousand (Hernandez 2012; see also Tuckman 2012a). With demands so massive, the Charter reads like the wish list for an urban utopia. According at least to the civil society membership of the CPCCMDC, this is precisely how it is supposed to function, as the highest aspirations that can be imagined, the perfection of Mexico City as "the city we dream of," as the first musings christened this vision (Zárate 2011).

The Charter, however, lacked any force of law. As soon as the public pomp and ceremonious circumstance subsided, the reason for the CPC-CMDC and the political energy that birthed and animated it (emanating from the mayor's office) evaporated. Without a public mandate, in other words, the coalition that had been the CPCCMDC transformed into an entirely different sort of political entity, one without any formal authority and with only the strength of their individual organizations and relationships to use as resources in pursuit of the right to the city.[30] Indeed, in the wake of the Charter's endorsement the right to the city became, or at the very least was more fully recognized by its advocates as a relational program, and the Charter a relational device. That is, in line with the critiques of rights talk stretching back to Marx noted above, it became apparent in Mexico City almost immediately after the Charter's endorsement that it

required the backing of significant force if it was to have any political purchase. Out of this realization came several arguments about how to proceed. Some coalition members posited the important reminder that the
Charter, though important, was intended by many (including Ebrard) as
a stepping stone on the path to a constitution. Others suggested that the
Charter itself should be given the force of law, and that its legal recognition
should be sought through the Legislative Assembly. Still others proposed
similar charters be sought at the level of individual delegaciones, with each
sub-municipal unit defining and legally establishing its own right to the
city priorities. Representing the GDF, Undersecretary of Government Juan
José García Ochoa synthesized several of these options into a three-pronged
approach: the construction of "a culture of the right to the city," "the modification of the legal framework" within which it exists, and the development of a constitution to "integrate and substantiate this new right" (García
Ochoa 2014, 12). Seven years later, the city would indeed have a constitution, though this advance would travel a different path than many in this
coalition intended (discussed in Chapter 5). Some work was also done on
creating borough-level charters, especially in Iztacalco.[31] In the main, however, pursuance of the right to the city after the Charter's public endorsement has drawn on relationships developed through the CPCCMDC and its
activities, rather than through the development of more forceful legislation.

Two prominent examples will help illustrate this general trend. The
first of these involved putting the Charter and its ideals to work by way of
demand, drawing on a relationship between coalition members and one
of the heads of the delegaciones that had signed their endorsement, Clara
Brugada Molina. Then a member of the PRD and executive head of the
expansive peripheral borough of Iztaplapa, Brugada had been an active
participant in the MUP before pursuing a career in public service. With
the support of the Ebrard administration and civil society partners, in 2012
Brugada completed a massive rehabilitation effort in the languishing Parque
Cuitláhuac, claiming the initiative as a way to extend the right to the city to
Iztapalapa's residents. Known for decades as the biggest garbage dump in
the area (Mora 2012), the former Santa Cruz Meyehualco landfill and surrounding areas were completely reconstructed and reimagined, with new
facilities and family friendly activities including leisure gardens, artificial
lakes, go-carts, and a zoo. The park's overhaul followed the ideals and principles established in the Charter with impressive breadth, incorporating
public participation in the construction and maintenance of certain areas
and pursuing such other ends as enjoyment, the social function of the city
and property, sustainability, and even the mitigation of environmental crises

ranging from global climate change to local air quality and the availability
of foodstuffs, all while consuming "very little" public funding, according to
Brugada (*GreenTV Noticias* 2012; see also Gómez Ayala 2011; Páramo 2012;
Quintero Morales 2012).[32] The 2012 inauguration of the park garnered the
attention of local and international news outlets, and city officials and local
residents heralded the arrival of this badly needed public space, calling it
the Park of the Rights of the People or the Park of the People (Parque de los
Derechos del Pueblo or Parque del Pueblo). Unfortunately, a good portion
of Parque Cuitláhuac's gleam wore off fairly quickly. Within months of its
opening the Legislative Assembly was publicly accusing borough officials
of mismanagement of the park's green spaces, which it was claimed were
being left to die despite the city's investment (ALDF 2013), among other
complaints.[33] Still, the undeniable success Brugada and her allies had in
turning forty hectares of retired landfill in one of the city's poorest and
most dangerous areas into a public park funded by the city remains one of
the examples cited by right to the city advocates as a successful instance of
putting the Charter affirmatively to work.[34]

The public backing of the Ebrard government also offered a different sort
of potential for putting the Charter to work negatively, by way of critique.
This potential was most clearly on display in the saga surrounding what
came to be called la Supervía Poniente (Western Superhighway, often simply Supervía) on the city's southwest side. Designed to connect the booming
edge city of Santa Fe, replete with immense hordes of international capital and glimmering new office and residential towers (some of them built
atop another old garbage dump and defunct sand mine) with the existing
ring road highway system known as the Periférico, the Supervía faced stern
residential resistance from its first official announcement. As the process of
expropriating land in the delegaciones of Magdalena Contreras and Álvaro
Obregón unfolded over the course of 2010, some residents found in the
Charter what they hoped would be a crucial tool of critique in their des-
perate attempts to halt the project. Having signed the Charter in July of that
year, local opponents attempted to hold Ebrard to account on precisely these
grounds, pointing to the contradiction seemingly at play in the mayor's pursuit and public praise of a project that residents and observers claimed
violated nearly all of the Charter's principles.[35] On one front, opponents
claimed the project would be destructive to the environment, principally
by contaminating and/or depleting ground water resources and contributing to the city's alarming levels of air pollution by encouraging car traffic
(Hernández León 2010). On another, they claimed the planning process had
violated the democratic rights of the citizens of the areas in question and

of the city more broadly. The project would use public resources to build a limited-access private superhighway the profits of which would accrue to a private partnership that would manage the completed project and the benefits of which would likely be enjoyed by less than one half of 1 percent of the city's daily auto commuters (Barros 2010).[36] Despite the obvious imbalance of affected parties and the scale of the project, city officials did not see fit to subject the project to the public scrutiny many felt was required by the Law of Citizen Participation (discussed in Chapter 4). Instead, the private partnership that controlled the project was allowed to conduct its own survey, which unsurprisingly reported a 74 percent rating of public favorability that was laughably out of sync with the increasingly organized resistance of local residents, academics, and professionals (Hernández León 2010; Salgado 2010).[37] Using Ebrard's endorsement of the Charter or the broader corpus of human rights as a mode of critique turned out to be a dangerous strategy, however. Acting on complaints it received from residents of the affected boroughs, the CDHDF conducted an extensive investigation into possible human rights violations associated with the project. Prior to the investigation's conclusion, it was rumored that Ebrard asked Commissioner Luis González Placencia to quash the report, which could be highly embarrassing to his administration and potentially to his presumed bid for the Mexican presidency in 2012. For its part, the CDHDF suggested a process of reconciliation take place between the Ebrard government and the formal opposition to the Supervía, el Frente Amplio contra la Supervía Poniente (hereafter FACSP), which the government refused (González Placencia 2011). In the absence of a reconciliation process, the CDHDF concluded and published a lengthy report in January of 2011, which cited six human rights violations associated with the Supervía committed by the city government and the two relevant delegaciones, and addressed particularly to the heads of government in each respective case, including Ebrard, whose name appears first in the report. Magdalena Contreras was cited for violation of the right to information and the right to citizen participation, Álvaro Obregón for violation of the right to legal certainty, and the GDF for those three plus the right to a clean environment, the right to water, and the right to adequate housing (CDHDF 2011). Allegedly perturbed by the publication of this report, Ebrard is said to have all but ensured that González Placencia would be relieved of his post by the incoming jefe.[38] In 2012, Miguel Ángel Mancera Espinosa did indeed appoint a new Commissioner of the CDHDF, Perla Gómez Gallardo. As Commissioner, Gómez has reportedly not enjoyed the friendliest of relations with civil society groups, but has maintained a very close working relationship with the Jefatura.[39]

And though its regular publication *Dfensor* has published several articles and special sections on the idea of the right to the city in the past several years, the CDHDF has under Gómez's guidance been less active in pursuing the right to the city than it had been under González Placencia. This difference has been interpreted as the loss of a significant institutional partner by leading figures in the coalition.

The two cases of Parque Cuitláhuac and the Supervía Poniente represent two avenues for pursuing the right to the city in the wake of the Charter's endorsement, one via demand and the other via critique. What the passage of time has illustrated in both cases is that the moment in which the right to the city enjoyed a privileged position as a discursive trump and the Charter a powerful political tool was all too fleeting, perhaps already fading even as it began to be realized by the coalition. The members of the CPCCMDC, after all, each had other institutional priorities that continue to draw their attention to matters of consequence outside the coalition. These groups continued to meet and discuss right to the city issues, and the phrase continues to find relevance in NGO and occasionally CDHDF reports, academic and public seminars and roundtables, and, somewhat surprisingly, the public remarks of Mexico City politicians, though in this last case the phrase seems safely relegated to those without aspirations for higher office. And even if the transition of the headship of the CDHDF had happened for reasons wholly unrelated to the Supervía Poniente and the criticism of Ebrard, the prevailing political winds from the mayor's office would seem nevertheless to have shifted away from the right to the city, a change of which the new CDHDF seems to have taken due note. Though Mayor Mancera was not openly hostile to the idea, his administration showed little to no interest in putting the right to the city explicitly on the agenda. As with Perla Gómez Gallardo, Mancera was frequently criticized for a seeming unwillingness to even take meetings with groups such as the MUP-CND or HIC-AL.[40] Each and all of these changes have had significant impacts on how the right to the city is pursued after the end of the political moment enjoyed by the CPCCMDC through the Charter's endorsement.

Mancera's lack of interest in the right to the city was a crucial nail in the Charter's political coffin. García Ochoa, the lone member of the Ebrard administration to retain significant authority in that of Mancera, disputed common characterizations of Mancera as disinterested, however (interview with the author November 26, 2015): "So it's not that it's less important, what happened is that the consensus that existed about instruments like this one . . . it collapsed." The Mancera government had also witnessed the Charter's use as simply a "banner against the city government" by certain groups, rather than an accord of mutual benefit, he further explained. Mancera was thus not opposed to the idea, but rather wished to build a new consensus around the right to the city if this idea was to find any relevance in his sexenio, which is precisely why, García explained, it was a theme to be explored at a public forum to take place in December of 2015. "So the [Mancera] government *is* interested in discussing this, but it is clear to us that it cannot only be used as a rallying cry [against the government], but rather must return to being an instrument of the government, and of [civil] society." Rather than a deep ideological break or some other sort of fundamental disagreement, García's interpretation of Mancera's stance in one sense merely points out the obvious, the common executive practice of disassociation with previous administrations. Mancera on this view appears an executive eager to forge his own path and avoid the pitfalls and unforced errors of Ebrard's final months as mayor, which not only tarnished his reputation but reportedly significantly dampened his chances at national elected office.[41] From a different vantage, however, García's explanation reveals again the dangers of dialectical inconstancy for instruments like the Charter. As argued above, any and all attempts to stabilize it as an operational concept, to fix its tentacles in place and secure from it a set of solid and dependable meanings, do a certain violence to its web of possibilities. In this case, the Charter addressed itself to a city that no longer exists, a city whose government has drastically changed; whose neighborhoods have witnessed demographic, aesthetic, and other transformations; whose daily rhythms no longer sync as tightly with the afternoon rains that come like clockwork in the summer or with the sleepy melodies of the organ grinders once found on every corner of the historic district; and whose place within the federation of Mexican states was about to be constitutionally realigned.

CONCLUSION: RETHINKING THE RIGHT TO MEXICO CITY

Building on the work of Walter Benjamin, Wright (1999) calls the Mexican woman in the border town of Ciudad Juárez a dialectical image, a set of material processes in a state of suspension, frozen in place and captured in only a momentary expression, as in a still life. Though the image at the center of this case is incredibly different, the fetishization of such stilled life is no less instructive. Despite the best efforts of CPCCMDC members to imbue their Charter with freedom of movement and to give it more than the narrow life the law often compels, its crystallization imposed limits on the creative process of becoming by which it was originally conceived. More importantly, the document itself and the political process that gave rise to it cemented a certain history with which the concept of the right to the city is now burdened. The history the Charter now carries is not, of course, universally burdensome, however. Indeed, favorable citation in assessing human rights cases by the CDHDF, public affirmation by prominent political figures, and broad popularity among powerful social movements and the academic Left in Mexico City endow this language with a certain amount of political capital. The point is not so much that the Charter froze forever the amorphous and fluctuating capacities of the right to the city in Mexico City, but rather that its fixity created a set of conditions with which any set of users and makers must now inevitably contend.

Remaining coalition stalwarts continue to promote the concept of the right to the city, forging connections whenever possible between this language and other conceptual apparatuses or material circumstances that might present themselves. As new urban struggles emerge, I was told, those involved make their own decisions about whether or not to engage with the right to the city and the complex relational entanglements it now entails. While its use may connote uncomfortable alliances or evoke the scars of bygone campaigns the memory of which may provoke certain foreclosures, it may just as easily create institutional access points or useful legal footholds. As a leading coalition member put it in 2015, "We offer this idea of right to the city to all of them. Some of them, the Supervía, decide to use it sometimes . . . and other times they didn't. . . . If they like, the concept is there. But for sure it's right to the city from our point of view."⁴² This attitude makes clear that even the formality of municipal participation cannot completely freeze an idea as unwieldy as the right to the city, whose ultimate potential resides in its dialectical relationality, flexibility, and conceptual and political fertility. In the conversation cited above, the issue then in question was that of the Corredor Cultural Chapultepec, the subject of the following chapter. Though never framed explicitly in the language of the right to the city, the opposition that formed against the Corredor project was informed and influenced by coalition members, many of whom shared the above sentiment that such politics belong to a struggle to achieve the right to the city whether or not their political advocates trade on this moniker. The matrix of conditions in which this choice must be made is in constant motion, as political figures and party factions move in and out of spotlights, as alliances are forged, broken, and reforged, and as dynamics that unfold at scales ranging from the neighborhood to the global economy push and pull on the priorities of the citizens of Mexico City. Some of these conditions change slowly, as in the case of the uneven, but steady spread of democratization in the capital city. Others happen at breakneck pace, as when the once-promising political star of Mayor Ebrard took an unexpected dive, and his seal of approval moved into an uncomfortable purgatory between blessing and curse. Amidst the maelstrom of changes now swirling around the citizens and militants of the Mexican capital, the right to the city is often made to travel and labor incognito, under any number of assumed names.

Choices as to whether, when, and how this language is taken up or taken on in Mexico City are therefore not academic or ideological so much as they are logistical and political. And in this respect the position of the right to the city in 2016–2017 was precisely where some leading members of the CPCCMDC intended that it should be, squarely—if understatedly— bolstering and perhaps enabling the push for the city's first constitution. After all, Jaime Rello Gómez, Marcelo Ebrard Casaubón, and Juan José García Ochoa, among many others, all publicly extolled the virtue of the Charter is precisely this capacity, and *La Jornada* announced its mayoral endorsement not as a triumph of a local charter of rights but rather as "another step toward the Federal District having a constitution" (Romero and Cruz 2010). In his speech on that fateful day, Ebrard made no attempt to hide his purpose in supporting the Charter, using the idea of the right to the city as a collective right not to claim a set of rights for the city's various communities, but rather to act as the claimant of collective rights that he argued ought properly to accrue to the city (as a collective subject) vis-à-vis the federal state. "The objective for this year," Ebrard (2010) intoned, "is that the Federal District will at last achieve its own constitution. . . . Two hundred years is sufficient time for the restitution of the city's rights, and it can determine its own constitution."[43] According not only to his own claims but to other CPCCMDC leaders, it was hoped that the Charter would form the basis of the city's new constitution, but the priority was always clearly understood in this line of thinking, that the Charter was merely a means to a constitutional end. The contests over and contours of that end, including the return of the right to the city to center stage of the urban revolution after its purported political death, are the subject of Chapter 5.

CHAPTER 4

Así No (Not Like This)

Resisting Postpolitics on Avenida Chapultepec

The task that someone effectively interested in political life might undertake far exceeds what is offered by the world of "politics." Politics ceased to be such because it is no longer based on reflective discourse, on a confrontation between proposals from society itself. **Bolivar Echeverría** (2019, 168))

It is in such a context that urban planners, experts, activists, and residents have invested urban infrastructures—from a renovated transportation hub to parking meters—with the power of producing a domesticated, beautiful city emptied of threatening bodies and social relations. Street workers in contemporary Mexico City are thus no longer material, visible signs of a future promise of inclusion and security. They appear as merely an out-of-place excess, a part with no part, in the civic collective. **Alejandra Leal Martínez (2020, 266)**

On the afternoon of Sunday, December 13, 2016, a small public square in the heart of Colonia Juárez was taken over by jovial celebration. Friends, neighbors, community organizers, and activists came together to laugh and socialize, and to pound piñatas crafted in the image of two powerful political figures whose planned megaproject on the edge of the neighborhood had recently been defeated by their collective efforts. They gathered to celebrate the victory of their No vote in the delegation-wide consultation that had taken place one week previous, a vote the GDF had unexpectedly decided to honor. Beyond the obvious joy of saving their neighborhood from the ravages of the project, many among the gathered also noted the symbolic and material significance of their victory to the future of local organizing and political power in the city. Before the vote, the prevailing

sentiment—even among those activists and local leaders firmly committed to the defeat of the project—was overwhelmingly pessimistic about
the role local voices could play in the political drama of the city's planning
processes. As one man asked me just days before the vote in response to a
question about its significance, "But what does it matter? They're just going
to go ahead with it anyway." After Mayor Miguel Ángel Mancera's unexpected public pronouncement following the vote that the project would
not go forward, however, the smallest remaining hope in the promises of
democracy began to grow, stoking the dwindling energy of el monstruo's
oft-disheartened grassroots.

The planned redevelopment of Avenida Chapultepec and its opposition
is part of a larger story of how the political climate in Mexico City came
to appear so dismally overcast to its residents, and how the clouds were
made to part again, if only for a moment. This episode therefore provides
an especially clear vantage on the decades-long struggle over the nature,
extent, and possibilities of democracy in the capital, and exemplifies both
Mexico City's trend toward what has been called the "postpolitical condition" (Swyngedouw 2011) and the moment the advance of this condition
was halted, even reversed. This moment was made possible by disparate
elements that coalesced as anti-project forces, themselves equally products
and productive of temporally and geographically specific conditions. In
this corner of space-time, the "suturing" processes that parse and divide the
social world into what Rancière (1994, 173) calls the "division of the percep-
tible" or the "partition of the sensible" was made to rupture and split, and
the frayed edges of its purposes were, for a time, made to serve unintended
ends. More significant than the empirical reality of this potentially fleeting victory, however, is the primary means by which it was accomplished.
The strategic vulnerability of the postpolitical condition then pertaining in
Mexico City came from within, from the hegemonic discursive machine
and its most effective rhetorical tool—the idea of democracy—and the legal
and political mechanisms promoted under its banner, which had together
produced increasingly stable control of municipal politics since the 1990s.
By shrouding their actions and purposes in the rhetoric of democatization
and seeking the legitimacy it provided, hegemonic actors also created the
conditions for the undoing of their well-laid plans along Avenida Chapultepec. In its aftermath, many residents began to hope that the unexpected
outcome of this struggle may hold a wider purchase for reworking democracy at the local level.

This chapter will examine the struggle over what was initially called
the Corredor Cultural Chapultepec-Zona Rosa (Chapultepec-Zona Rosa

Cultural Corridor), beginning with a brief reading of some of the specific political conditions pertaining in the city before the vote, as a targeted addendum to the elaborations of Chapters 1 and 2. Reading this history through the lens of the postpolitical, the following section also outlines salient critiques of this framework and argues for a more nuanced approach tailored to specificities attendant to postpolitical instantiations in particular contexts. The third section describes how opposition formed quick on the heels of the project's public rollout, and details some of the factors that allowed for its ultimate success. The concluding section further elucidates the two central claims of this analysis, that postpolitics is operative and must be combatted at the level of quotidian norms, assumptions, and attitudes, and that resistance to postpolitical orders can be effectively pursued through strategic vulnerabilities located in the spatio-temporal conditions of emergence pertaining to their specific hegemonies. This analysis contributes to the broader contemporary imperative "to rethink urban politics and urban political theory in ways that are much more sensitive to considering the city as an imminent site for nurturing political subjectivation, mediating encounter, staging interruption and experimentally producing new forms of democratization," especially in light of the globally extant trend of urban depoliticization (Dikeç and Swyngedouw 2017, 3).

APPROACHING MEXICAN POSTPOLITICS

PRD governance of the capital city was something many in the city had long hoped and worked to make possible, and represented, at least for some, a respite from PRI's history of electoral manipulation, graft, intimidation, and violence.[1] From its very inception, the PRI had sought to quell dissensus through incorporation, and had produced a measure of legitimacy through its roughly seven decades of continuous and unassailable electoral victories.[2] With the 2012 return of the PRI to the national executive, some feared a return of the corruption and intimidation they associated with previous PRI regimes, along with the reactivation of party machinery lying in wait for the party's return to presidential power. Both possibilities portended a kind of death for the great hope of democratization many commentators and residents had seen in the 2000 election.[3] The PRD, too, had trouble keeping the faith in its stewardship of the capital. Several of the regimes after Cárdenas's departure from the Jefetura were mired in allegations of corruption and a marked lack of transparency, and many argued that the party's manipulative and clientelist tactics bore an unsavory resemblance to those long associated with the PRI.[4] As the city's democratic alternative, the party was therefore already faltering even in its fledgling years, and showed little evidence of making good on its promises of meaningful democratization. Instead, the party had set to work cultivating and exercising a political hegemony much like that of the leviathan it had sworn to defeat. The rather shameful irony of this, while painfully resonant for many contemporary *vecinos*, seemed lost on their leadership. In a stinging critique of this widely agreed-upon failure poetically drawn from the PRD's own hopeful nickname, Gugelberger (2005, 107) dourly reported in 2005, "We now speak of the setting of the *Sol Azteca* (the PRD)."[5]

The Ley de Participación Ciudadana (Law of Citizen Participation, hereafter LPC) is the main measure through and by which the PRD and the GDF promised and pursued democratization, and also an important instrument through which party leaders consolidated and exercised postpolitical hegemony over the urban planning process. Essentially, the LPC was ostensibly designed to codify chilangos' right to participate, in various ways, in the process of pursuing major public initiatives. First approved by the Legislative Assembly in 1995, the law provides twelve instruments for organizing citizen participation (since 2010 reforms), including plebiscites, referendums, consultations, and citizen assemblies (IEDF 2016).[6] While some of these measures allow citizen participants to deliver results that carry the force of law over various authorities (such as the mayor's office or the ALDF), most do not.[7] Despite the proliferation of such legal implements for enabling citizen participation, however, Harbers (2007) argues that such mechanisms serve the aims of regime/system legitimation and charismatic, neopopulist dramatics rather than meaningful participation, building leaders like AMLO into seemingly infallible figures that hearken back to the caudillos of Mexican history. This skepticism of Mexican participatory schemes also extends far beyond the LPC, through numerous local and national PRI regimes in which legitimacy and pacification were achieved through programs and projects that rarely extended beyond nominal citizen input.[8] Still, democratization and ever-greater participation are the ideas by which the PRD sought to establish its control of the city, especially in the post-transition years. The party of Cárdenas continues to trade on its opposition identity and democratic ideals. Even its manipulation of the planning process, exemplified by the LPC, is couched in the languages of democracy and participation.

Given this history, it is not difficult to understand why the city feels rife with political cynicism, a quotidian theme so recurrent in daily conversations as to be rivaled in my fieldnotes only by feelings of political disappointment and apathy.[9] From this vantage, such disillusionment with the government could easily slip analytically into the well-worn ruts still cutting through scholarship on Latin America, such as "the norm of illegitimacy" in government (Horowitz 1969) and various other takes on everyday corruption, or side-eyed appraisals of the gaping distance observed between the peculiar pageantry of Mexican politics, with all its brazen pomp, and the actual, behind-the-scenes deals and dictates of its daily animation.[10] Rather, what this evidence suggests is the hegemony of a particular set of mechanisms, practices, and attitudes often described as the onset of a postpolitical condition.[11] That the roots of such a condition in Mexico City should be found in the PRI regimes previous to the 2000 transition is likewise no surprise at all, given the open secret of the party's uses of democracy. Indeed, quite aside from Vargas Llosa's infamously incendiary remarks, the realities of the Mexican state and its municipal counterpart have long betrayed the blurring of the boundary Rancière (2007, 18–19) notes between "good democracy" and its tyrannical "other":

> Does not the best of democracies, indeed the good *politeia*, where the mass of citizens fulfil [*sic*] their preference for lucrative activity over the activity of citizenship, in short that good political regime which coincides with the satisfaction of citizens' apolitical needs, bring into play the very same mechanisms which serve the tyrannical annihilation of collective power: *microphronein*, the smallmindedness of individuals locked into pettiness, the idiocy of private interests; and *adunamia*, the impotence of those who have lost the resource of collective action? Smallmindedness, mistrust and the impotence of the citizens—these are the means of tyranny, all the more liable to resemble the means of good government.

Residents of Mexico City have long participated in a political system—whether under the control of the PRI, PAN, or PRD—dancing on both sides of this boundary, with parties and presidents availing themselves of the weapons of tyranny while shrouded in the garb of democracy, which provides, as Brown (2011, 57) puts it, "a gloss of legitimacy for its inversion." The sobering reality daily reproduced by the mass of chilangos thus echoes the deceptively simple gauntlet thrown down by Cruikshank (1999, 2), building on the words of Foucault, that "'everything is dangerous,' even democracy."

By the time of Miguel Ángel Mancera's assumption of power in 2012, these developments in the political landscape of chilangolandia had coalesced into a state of affairs most productively viewed through the lens of postpolitics. This postpolitical condition describes a state of relations in which dissensus is viewed and treated as beyond unproductive, as something unreasonable. As in the related frame of post-democracy,[12] this condition instead takes on the ethos of consensus, wherein the greatest good is achieved when parties (or stakeholders) work within a set of given norms toward an agreed-upon set of possible ends. Wilson and Swyngedouw (2014, 6) argue that across diverse conceptualizations of postpolitics, "they all refer to a situation in which the political—understood as a space of contestation and agonistic engagement—is increasingly colonised by politics—understood as technocratic mechanisms and consensual procedures that operate within an unquestioned framework of representative democracy, free market economics, and cosmopolitan liberalism." The key is the removal of fundamental disagreement from the stage of politics, with all counted parties continually reaffirming the boundaries of acceptable discussion and eschewing as irrelevant, counterproductive, or dangerous any and all potentially destabilizing elements. Any radical potential, anything that could undermine or otherwise threaten the terms of debate, must be foreclosed. This careful management of political discourse is what Rancière calls the "partition of the sensible," the framework within which utterance is read as either noise or speech, and action as acceptable or disruptive. The specific form this partitioning takes in a given society is known as the police, or the police order.[13] The police order, as Dikeç (2005, 174, original emphasis) explains, "refers to an established social order of governance with everyone in their 'proper' place in the seemingly natural order of things," an order "achieved through the configuration of a perceptive field, through the symbolic constitution of the social, which becomes, from the viewpoint of the police, *the* basis for government." Politics, true dissensus, has already retired from the scene inside a given police order, hence the characterization of such orders as postpolitical. The practical options within such a scenario have often been characterized as having been whittled down to anemic participation in the agreed-upon structures bound to the police order, which brings tacit complicity in that order and is anyway fundamentally incapable of remedying its perceived systematic ills, or similarly powerless outbursts of violence or indignant abstention from accepted political behavior.[14]

As an analytical framework, the postpolitical thesis has run afoul of critics on several grounds, not least in Anglophone geography.[15] McCarthy (2013) points to elements of eurocentricity at work in the development of the very idea of postpolitics, countering Swyngedouw's (2009) argument that the non-politics of climate present a clear material example of the postpolitical condition by contrasting Swyngedouw's observations with the white-hot reality of the US political debates daily waged over climate, from the increasingly garish denials of presidential candidates to the hard-fought battles over lightbulb regulations. While the postpolitical thesis may hold in some contexts, this critique suggests, its global reach even over its most comfortable thematic terrain has been overstated. Hannah (2016) extends this line of argument in characterizing postpolitics as a "distinction-collapsing discourse" that serves in practice to feed the specter of "state-phobia" by too easily dismissing the state as a consequential relational field of contest. Mitchell, Staeheli, and Attoh (2015) likewise argue that in following Rancière into a strict notion of "proper politics," the postpolitical literature has been unwilling to recognize as political those many actions without which a given police order could not be sustained, or those by which it is initially and continuously (re)constituted. Similar to the insightful prod at the academic Left propelled by McCarthy's (2002) exposition of the Wise Use Movement, Mitchell, Staeheli, and Attoh (2015, 2645) caution that the dismissal of such activities threatens to "equate politics that we do not appreciate for an absence of politics." Davidson and Iveson (2015, 551) echo this caution, but also counter that Rancière's work "has sought not to define and/or identify a preferred political agent or place," rejecting "any notion that particular places and/or people are the proper spaces and/or subjects of politics."

That postpolitical theorizing should fall short of critical expectations on the issue of resistance in particular is troubling. Part of the problem of course lies in the distinction collapsing by which the multiple variants of postpolitical theorizing are reduced to a singular thesis.[16] As Wilson and Swyngedouw's (2014) useful schematic indicates, the three most significant theorists of postpolitics—Jacques Rancière, Chantal Mouffe, and Slavoj Žižek—differ markedly in their understanding of the constitution of the political and of politics, in the strategic orientation of their critiques, and much else. Approaching this conceptual terrain through the generative differences highlighted by this schematic opens up space for strategic recombinations of this analytical scaffolding in order to more precisely and productively understand resistance to postpolitical instantiations in situ. For Rancière, while the properly political act or sequence may be rare," it always remains an imminent possibility, as the police order can never suture or place the social body in its entirety. "There is always a gap, a void, a lack, or excess that resists symbolization, a hard kernel that is not accounted for in the symbolic order," Swyngedouw (2011, 374) argues, a remainder that "stands as guarantee for the return of the political." Political gestures, instigated by the surplus element of the police order as assertions of the "part of those without part," however, do not come from some party or entity a

priori deemed outside the policed perceptive field, but rather "reject existing identifications through a process of political subjectification that gen-
erates identities outside of the existing police order" (Davidson and Iveson
2015, 548). Such moments are both generative of alternative formations and
destructive to those of an established perceptive field, a violent process "in
which bodies are torn from their assigned places, and exhibit verbal com-
petences and emotional capabilities they are not supposed to have by virtue of the space-time they occupy" (Corcoran 2010). Indeed, the police
order itself creates the basis for such identificatory transgression. As Dikeç
(2005, 181) argues, "If politics puts the police ordering of space to an egalitarian test, then politics is possible not despite the police, but because of
it." Rancière's approach thus provides a general language for conceptualiz-
ing resistance to a police order and the ways in which such orders attempt
to subdue, suppress, or preferably incorporate and by extension implicate
potentially radical elements through the suturing of the social body. And
though the particular qualities of the omnipresent radical remainder they
envision cannot be known in advance—in belonging, as Derickson (2017)
argues Rancière's project does, to the realm of ontology—their emphasis on
its political potential nevertheless provides a useful analytical posture from
which to understand resistance.

Grounded in a vision of antagonism rather than egalitarianism as the
basis of the political, Mouffe's explicitly hostile approach to postpolitics conceptually provides far more specificity by emphasizing the spatio-temporal
contingency of such political orders. Equally important, Mouffe provides a
critique of postpolitics which distinguishes it from a broader field of depoliticizing orders by emphasizing a postpolitical tendency toward legitimation through the nominal furtherance of liberal democracy. Much of *On
the Political* is addressed to a self-celebratory postpolitics that lauds liberal
democracy's besting of all significant ideological foes, leaving only backward fundamentalists huddled angrily in the few remaining dark corners
of the global political map, and that proclaims its era beyond hegemony. In
attempting to disabuse the world of the notion that any such victory over
political antagonism is verifiably extant, Mouffe (2005, 18) instead views
the postpolitical as being subject to fundamentally the same processes of
instantiation and contest as other political orders, a "temporary and precarious articulation of contingent practices." A given police order, on this
view, is merely an established "*hegemonic relation*" (Laclau and Mouffe
2001, xiii, original emphasis), in which "a certain particularity assumes the
representation of a universality entirely incommensurable with it." Postpolitical police orderings of the perceptive field are thus productively viewed

as "attempts to impose a construction of unity on the real ground of difference, or a construction of agreement on the real ground of antagonism"
(Purcell 2013, 563), and therefore entail an obfuscation, denial, or "repression" of their "constitutive outside." Regardless of whether a new hegemony
(another police ordering) would substantiate an agonistic model of politics
or merely a different sort of oppressive leviathan equally bent on the saturation of the social body, Mouffe's approach insists on the empirical specificity of the counter-hegemonic resistance by which it must be pursued. To
understand these processes as hegemonic means to conceive of postpolitics
not as a global condition that, once established, need only deal with occasional violent outbursts that only cause minor irritations to its equilibrium,
but rather in its multiplicity as political projects always in need of reproduction and always susceptible to attacks that strike at their very foundations.
Any postpolitical order, this orientation suggests, is historically and geographically specific, a political victory achieved and contested in a particular place-moment. It is an inherently unstable political fix, the very terms
of its construction and maintenance in constant danger of the subjective
violence always lurking on the underside of its plane of imminence.

It is to this foundational terrain that researchers must look in seeking to
"recover the political."[17] Each and every instantiation of postpolitics must
be deeply rooted in its peculiar spatial and temporal specificities, its suturing of the perceptive field anchored in the wounds left on the landscape by
material and social forces alike. The postpolitical order of recent years in
Mexico City, most saliently operative through the use of the LPC, provides
an ideal empirical context for understanding resistance to such orders and
similarly de-politicizing regimes of governance.[18] Specifically, analysis of
this resistance will demonstrate that strategic vulnerabilities of such orders
can be fruitfully sought in their conditions of emergence and the politics
by which their hegemonies are initially constructed and their postpolitical architectures erected. In Mexico City, the long night of the embattled
PRI called forth the figure of the democratic transition, and democratization became the tool through and by which the PRD took control of the
municipal—and the PAN the federal—state. When the promises of democratization lavishly lauded from within and without yielded only populist
pageantry and the proliferation of participatory planning mechanisms of
the sort derisively regarded as cases in point by the postpolitical literature, fatigue, apathy, and cynicism took hold of a significant share of residents as the reality of a postpolitical and postdemocratic Mexico City came
starkly into view.[19] Rather than continue to accept the observed reality that
democracy functions only as a tool of legitimation, as a rhetorical weapon

cheekily brandished by political elites in their public dances of opposition but secretly wielded against a sleepwalking *demos*, opponents of the Corredor Cultural Chapultepec found in the concept of democratic participation a cornerstone of Mexico City's contemporary postpolitical reality, the lynchpin of its policed perceptive field. Rather than engaging what was largely considered a flawed (at best) or fraudulent (at worst) democratic process the end result of which was nearly universally assumed, then, resistance focused on this strategic strike point in the police order, its most foundational organizing principle. Their unexpected victory caused a tear so deep that ordinary citizens in some of the city's oldest neighborhoods exhibited genuine shock in the aftermath of the December 6 vote. To accomplish this result, resistance first had to combat the oft-cited apathy of local residents and the feelings of disconnection, cynicism, and hopelessness that pervade their appraisals of the political process, a fundamental dimension of postpolitical hegemony rarely engaged in the literature. Analysis of the multivalent quotidian expressions and tactics of this militancy, therefore, also answers recent calls to look beyond narrow understandings of proper politics in the analysis of postpolitical contexts, widening understanding of counter-hegemonic action and focusing attention on the everyday, subjective operation of postpolitical hegemony outside the realm of institutions and policy.[20] Residents waded cautiously into their victory with wide eyes, spreading out to ask one another if it was real, or too good to be true. As caution gave way to joy at their victory, anger at what had been their powerlessness was forced to turn toward a generative outlet. "I always say," architect and Juárez neighborhood activist Sergio González announced at a public forum on April 26, 2016, "that Mexico City will never be the same after December 6." The following section explores how a postpolitical hegemony was contested in Delegacion Cuauhtémoc during the latter part of 2015.

ENACTING MEXICAN POSTPOLITICS: THE CORREDOR CULTURAL CHAPULTEPEC

The impetus for some type of improvement on Avenida Chapultepec has been around for some time, and murmurings that the city was planning some kind of project there did not come as any great shock to many local residents. The surrounding neighborhoods had been slowly changing for several decades in a way that made the avenue stand out aesthetically—and not in a good way. The rationale the public/private development agency ProCDMX and its head (and prominent public ambassador), Simón Levy, offered for the planned intervention would likewise have been unsurprising

FIGURE 4.1. A map of Avenida Chapultepec and surrounding area, including the site of the proposed Corredor project. Cartography by author.

FIGURE 4.2. Paseo de la Reforma from the Ángel de la Independencia. Photography by author.

FIGURE 4.3. Avenida Chapultepec near the Sevilla Metro station. Photography by author.

to even a casual observer of such improvement projects. The avenue, the GDF argued, had become dangerous, run-down, both under- and overutilized, an obstacle to progress. It thus presented an opportunity to "rethink the city" for automobiles, cyclists, and pedestrians, Simón Levy mused via Twitter, in the months leading up to the project's unveiling, later ethereally adding: "Avenida Chapultepec is not the renewal of an avenue, it is urban consciousness beginning with social vision."[21] Exploring the specificities of this terrain as then constituted can reveal much about precisely how the "social" was here envisioned, which parts of the social this perspective was unable or unwilling to capture, and why the city needed such rethinking.

Avenida Chapultepec (see Figure 4.1) runs east/northeast from the northeast corner of Bosque de Chapultepec to the city's historic center. Though they increasingly diverge as they move away from the park, the avenue runs roughly parallel to Paseo de la Reforma, the city's most historically significant boulevard, which also runs from the Bosque to the historic center (see Figures 4.1 and 4.2). Reforma, as it is often called, has a long and turbulent history extending farther back even than the Revolution. Its name commemorates the reforms of (mostly) beloved Indian President Benito Juárez, the great liberal modernizer. Reforma was envisioned by Napoleonic Emperor Maximiliano of the Second Mexican Empire as a promenade to rival those of the great European cities, and as a swift and aesthetically pleasing route between the city's government buildings in the historic center and his preferred palatial residence atop the ramparts of Chapultepec Castle.[22] Lined with trees and littered with small monuments and memorial busts of legendary figures, Reforma passes through several *glorietas* (traffic circles) encircling monuments and gardens, including an infamously sensual statue of Greek goddess Diana the Huntress, and the Angel of Independence, a site of national significance and a preferred locus of civic activity, especially social protest.[23] In recent decades, rounds of development fueled in large part by foreign direct investment, have taken Reforma from a sleepy historic promenade to a thriving business corridor catering to the dictates of the real estate sector and the desires of a transnational managerial class increasingly at home along its edges.[24] Visually, Reforma is now dominated by skyscrapers of heights unknown in Mexico until recent years, tattooed with the monikers of international finance and other transnational entities (see Figure 4.2). Avenida Chapultepec, however, has lagged behind its famous neighbor (see Figure 4.3).

Between Reforma and Avenida Chapultepec lies the colonia of Juárez. The section of Juárez closest to Bosque de Chapultepec is known as the Zona Rosa (pink zone), an area famously frequented by tourists and also well known for its gay clubs, Asian markets (especially Korean), and its youth and alternative cultures. It is a spectacular place, a tangle of groups and images uneasily swirling around each other, interspersed with bars, sex shops, US-based fast food chains, beggars, magicians, police, and *godines*. This last group, lightly pejoratively known by this term, comprises low-level functionaries of corporate and governmental enterprise, unmistakable in dress and demeanor and forever found in the cantinas, bars, and *loncherias* of Juárez, as well as the food carts and public benches that line Reforma and Avenida Chapultepec. They are the "sovereigns of sadness" Bolaño (2007) depicts in their zombie-like drudgery, working their days and drinking their evenings away in neighborhoods far too central for their meager wages. The rest of Juárez has a decidedly residential character, save for the slivers of commerce along Reforma and Avenida Chapultepec, the neighborhood market, and the small repair shops, convenience stores, cafes, and the like that occasionally inhabit the ground floor. Across Avenida Chapultepec to the south lie the neighborhoods of Roma (split into Roma Norte [North] and Roma Sur [South]) and Condesa (including Hipodromo and Hipodromo-Condesa). Condesa, centered on two lovely public parks (Parque España and Parque México), has become one of the city's more desirable neighborhoods in recent decades, and is favored by the transnational portion of the white-collar labor force employed along

FIGURE 4.4. The site of the tianguis above Chapultepec Metro station, March 19, 2016. Photography by author.

Reforma.²⁵ According to local residents and academic observers, Condesa is also an epicenter of gentrification in Mexico City. Though it boasts a thriving nightlife scene, Condesa caters to a decidedly wealthier crowd than does the Zona Rosa, and is invariably suggested as the premier area for such high order goods and services (and harbingers of gentrification) as fancy barber shops, romantic rooftop martinis, and Sunday brunches cute enough to make the hippest Brooklyn Instagrammers opine their jealousy. Condesa is also, however, becoming rather too settled for many younger residents, who increasingly find their pleasure profiles more suited to neighboring Roma Norte, one gentrifying wave from Condesa. Here, fancy art galleries and a new generation of globally recognized upscale cocktail bars mingle with dilapidated commercial and residential spaces in an aesthetic union distinctly chilango, but somehow also eerily familiar to certain pockets of gentrifying Chicago or New York.

In all three neighborhoods, the Porfirian grandeur of the architecture, abundance of well-maintained public space, reputed relative security, and proximity to the business corridor along Reforma have conspired to attract an ever-wealthier, younger, and more international set of users and residents. As the scene shifts from the overplayed Condesa to the far northeast reaches of Juárez, longtime residents have become increasingly uncomfortable with the presence of persons and land uses unfamiliar in their

FIGURE 4.5. The site of the tianguis above Chapultepec Metro station, May 29, 2016. Photography by author.

neighborhoods. The following scene, which unfolded at a community meeting in precisely this section of Juárez, captures this sentiment. Herón, a community organizer and activist, was attempting to explain the concept of gentrification to a meeting of local residents (fieldnotes, April 24, 2016):

> Thinking a concrete example the best approach, he suddenly changed track. "It's like this *bar clandestino* down the block." "What bar?" several people asked at once, with surprise. "It's right down the block, just one door from the corner," he began. "You mean the taco shop?," someone asked. "Ah, no, there's a secret bar *inside* the taco shop. You have to have a special membership, and they let you in through a secret false refrigerator door in the back. Inside, there's a fancy bar." The whole room was taken aback. They were scandalized. People audibly gasped, and I saw a woman cover her mouth with her hands. These neighbors could not believe that such a thing had come to their beloved Juárez. A secret bar for fucking *fresas* and *mirreyes*, right here in Juárez. What had the world come to? This gentrification had to be stopped.
>
> I had heard of this bar before. It was all the rage among the party set. Iggy had been angling for a free membership there for six weeks (the annual fee is something like 2,500 pesos). Alex and Mario were already members, as was Claire. Right there in the midst of the scandalized Juárez neighborhood school of citizenship, while the horrified *vecinos* consoled one another

and tried to make sense of this brave new world they hadn't known until this very moment they were already living in, I sent a text to Alex and asked if she could take me to this clandestine bar everyone was talking about. She said she would love to, and that we could go on Thursday night. I couldn't wait.

My own comical enthusiasm aside, this is precisely the kind of development city governments have privileged for several decades by encouraging the inflow of foreign capital and the subsidized improvement of its preferred landing zones (Parnreiter 2015). Seen from this angle, Avenida Chapultepec begins to look more and more like an eyesore, something too Indigenous, an impediment to the integration of the trendy neighborhoods pushing in on it from all sides, squeezing its informalities into smaller and smaller corners until, with no fanfare or notice at all, the police come and unceremoniously clear out the last vestiges, as happened to the longstanding *tianguis* located at the avenue's southwest terminus above the Chapultepec Metro station in the closing weeks of May, 2016 (see Figures 4.4 and 4.5).[26] Simón Levy explicitly noted such uses of public space as one of the crucial factors necessitating the agency's intervention, along with urgently needed neighborhood integration. In an interview with television station Teleforo in September of 2015, Levy explained that persons working along Reforma are beginning to live there too, and need to be able to cross the avenue more easily to access nearby neighborhoods. Moreover, developing retail potential along the avenue (by introducing the kinds of upscale eateries and shops then more at home in nearby Polanco) would keep this population's spending in the immediate area. Redeveloping the avenue, in other words, was necessary to connect the commerce of Reforma with the consumption and residential areas in Roma and Condesa, creating an insular orbit for work, leisure, and retail catering especially to the city's growing (and increasingly international) professional and managerial classes.

As then constituted, the avenue did, as proponents of the planned redevelopment claimed, present a significant danger to its many daily users. Level with sidewalks and medians in many places and with few designated crossings, the avenue's many lanes are dangerous to cross outside of crosswalks, and even sometimes within them (see Figure 4.6). As a thoroughfare connecting the historic center to the park and the fancy residential and commercial areas that surround it, the avenue continues, along with Reforma, to enable the routinized Maximilian escape from downtown, and from the city's lone international airport. This ensures that Avenida Chapultepec is full of reasonably high-speed traffic at nearly all hours, and most pedestrian crossers appear unsurprised by frequent close calls, not to

FIGURE 4.6. Avenida Chapultepec near the Sevilla Metro station. Photography by author.

mention cat calls, abusive language, and obscene gestures. At least as much as ensuring the safety of pedestrians, however, many users of the avenue wished planners to focus on the efficient flow of traffic, which is slowed by unauthorized crossers and street vendors (whose operations often spill over the sidewalk and into the edges of the street), as well as other unplanned obstructions.

A last set of factors not often discussed by the project's proponents (or not at all, as the case may be) concern the profitability and development potential the project promised for its backers, for the city as a whole, and for those functionaries who spend at least some of their time opening envelopes in the many dark corners of the local bureaucracy. Publicly, there was the occasional mention of the profitability of the project, though its aesthetic and cultural appeal were the real selling points. On another level, however, it remains an open secret that new leases and licenses present innumerable opportunities for graft. The specific local flavor of this system of extralegal financial arrangements appears, anecdotally, to have changed only in its facade under PRD control, in line with governing trends outlined above. Tread carefully enough, and some chilangos and international observers will cautiously (and often vaguely) relate the informed suspicion that such projects also present excellent opportunities for laundering vast amounts of money accumulated through narcotics and other nefarious enterprises, especially through unaccounted-for injections of cash that are said to magically appear to counter what observers skeptically view as all-too-expected cost-overruns.[27]

This convergence of factors is what led the city's planners to see and frame the improvement of Avenida Chapultepec as a necessity, an urgent demand transcending and superseding the petty squabbles and disagreements of local residents. It was a problem the city's planning elite should have been well positioned to solve according to their own social vision, and what was assumed to be the unassailable machinery of the pospolitical planning infrastructure centered on the LPC, and this is precisely what they set out to do in 2015.[28] By the time Simón Levy and ProCDMX unveiled the plans for the project in August, contracts had reportedly already been signed for its financing and construction.[29] Because democratization was the vehicle by which their postpolitical consensus had been achieved, however, the nominally democratic mechanisms of participation figured as an all-important last box to check in the realization of their plan. This should, by all accounts, have presented no problem at all, as the LPC contains several useful workarounds. The mayor needed only choose one of the several non-binding mechanisms for citizen participation, jump through the proverbial hoops, make a boilerplate public statement emphasizing the impact of this valuable input on the final judgement, and then proceed with the project otherwise unimpeded. Choosing the consulta option, wherein a non-binding vote would be framed in a simple yes/no question as to whether or not the potentially affected public favored the initiative—after several public fora in which residents could ask questions and voice concerns about the project—was the obvious first step down this path.[30] This process, however, unfolded in a manner decidedly less straightforward than initially predicted.

CONTESTING MEXICAN POSTPOLITICS

On August 18, Simón Levy and ProCDMX presented the finished plans for the Corredor Cultural Chapultepec-Zona Rosa, which had evidently been in the works in some form since the early months of the year. A detailed account of the planning process (bidding schedule, financing structure, project scope, some diagrammatic explanations of form, etc.) appeared in that day's *Gaceta Oficial* (the city's official gazette), and plans and scale models began to appear on tv news and across the Twitterverse. It was announced that there would be a public consultation process in which residents would be invited to comment on the plans and voice any concerns, a process to be carried out through public fora and surveys conducted by

FIGURE 4.7. Signs produced by anti-project forces, displayed in Juárez, December 13, 2015–December 15, 2015. Photography by author.

the IEDF. Another key figure of this public rollout was Fernando Romero, the architect who had designed the project. Romero, the son-in-law of Carlos Slim, also designed the retail and real estate titan's Museo Soumaya.[31] Resplendent by design, the art museum's outer shell—a tessellated mesh of aluminum hexagons—pinches suggestively between rounded rectangles top and bottom and reflects the light of the sun during the day and the glow of the city's own lights by night. The museum's futuristic form looms above the Plaza Carso at the edge of lavish Polanco, making an aggressively global aesthetic statement of the sort many feel abstracts away from the city's history and culture.[32] As is increasingly prevalent in prominent architecture in the city's upscale neighborhoods—especially Santa Fe and other development corridors created through the highly centralized planning practices discussed by Parnreiter (2015)—the Corredor was described in precisely such a globally aspiring frame, with Romero and Levy both making frequent explicit reference to New York City's elevated lineal park, the High Line. Like the High Line, the Corredor would contain a carefully curated array of Indigenous flora to be appreciated by orderly visitors within a small range of carefully scripted consumption and passive recreation possibilities. Also like the High Line, a large portion of the project would be elevated above the street, with sections of the street below appropriated for commercial, pedestrian, and bicycle use. The remaining roadway would also have to concede one lane specially reserved for a new Metrobus line.[33] The lineal park would culminate in a sort of amphitheater rising several levels above the Glorieta Insurgentes, where a large screen could be used to show videos.[34] Viewed as a scale model, the stark-white project loosely resembled a massive ocean liner, and some locals were soon referring to it as *el titanic*.

Initial distaste for the plans in particular and the planning process in general quickly began to coalesce into organized resistance. Architects and urbanists across the city criticized the project's design, citing especially its outward-looking character and what some saw as its blatant attempt to appeal specifically to expatriate professionals and wealthy Mexicans, quite aside from the concerns of gentrification and its likely corollary, displacement, and the ironic argument that elevated lineal features like this one were more likely to separate than to integrate neighborhoods.[35] Many also reacted strongly to the transformation of public into private space inherent in the project and the lack of citizen input in the development of the plans. After all, many pointed out, residents were being asked only to comment on highly detailed finished plans, rather than having input into the planning process itself. Levy had tried to avoid the privatization critique—quite predictable in Mexico City, given ongoing efforts to privatize everything from oil leases to communal farms and even communally held urban properties throughout Mexico since at least the 1980s—by structuring ProCDMX's control of the avenue through a renewable forty-year lease arrangement. This maneuver seems neither to have fooled nor satisfied most opponents of privatization in the city.[36] Cutting across these many critiques was the somewhat obvious superficial fallacy that while it was billed as a cultural corridor, the only culture on offer seemed to be retailers and eateries. For many detractors, this made the project little more than a fancy mall. This last critique held a special rhetorical sting, and also found resonance with a growing recognition of the commercialization of the GDF's cultural policy agenda under the PRD, especially after Cárdenas. "With public-private partnerships assuming an increasingly pivotal role in the shaping of public culture in the city," Kanai and Ortega-Alcázar (2009, 488) explain, "some argue that private actors increasingly appropriate the built environment for their own interest and exert undue influence on public policies." Confronted with these criticisms, Levy and Mancera continually reaffirmed their faith in the consulta process.

Rather than abstain from the process en masse or react with physical violence, as might have been expected, the area's increasingly organized residents responded directly on the terms provided by the planning process, however fundamentally flawed and however much a democratic sham they understood the consulta to be. Bombarding Levy with difficult questions

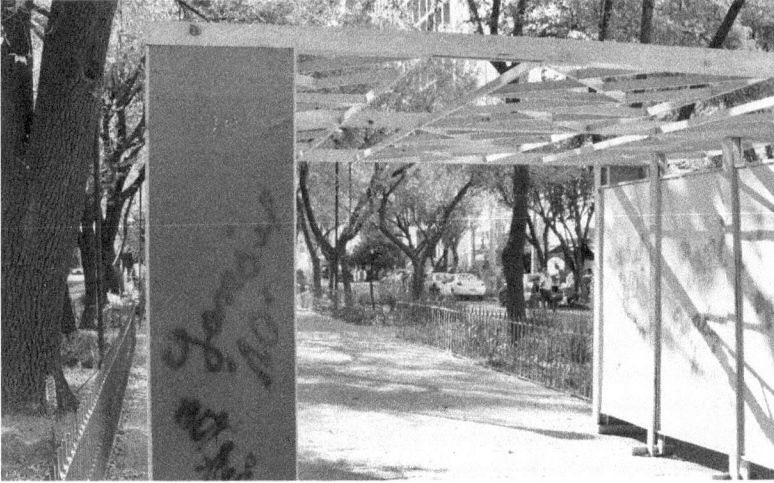

FIGURE 4.8. A structure promoting the Corredor Cultural Chapultepec vandalized with counter-messaging, still standing in Roma Norte some ten days after December 6, 2015. Photography by author.

and harsh critiques at public meetings, residents sought to have their dissatisfactions with the project heard and respected. It quickly became clear that most residents were not interested in adjusting the project, but only in rejecting it outright. When the Mancera government announced on November 6 that the consulta process would culminate in a yes/no vote exactly one month later on December 6, the battle lines were firmly drawn. Residents from Condesa, Roma, and Juárez held several marches along the avenue in the ensuing weeks, in addition to an aggressive media campaign.[37] Fliers, posters, handbills, and the like were posted on utility poles, across public works, and in the windows and across the balconies of residential and commercial spaces (see Figure 4.7). Facebook groups formed to promote resident criticisms of the project, and Twitter became a virtual battleground for promoters and critics alike. Local television carried images of the plans, interviews with and comments from Levy and ProCDMX, and reminders of the date of the vote. One of the more interesting and effective interventions was a YouTube video produced by *Los Supercívicos*, a group whose comical admonishments and profoundly uncomfortable shaming of poorly behaving residents, vendors, and city officials afford them some measure of local celebrity. In a five-minute-thirty-nine-second video, *Los Supercívicos* pulled no punches in their attack on Mancera and Levy's plans and their postpolitical planning process. In the video, when the residents of Juárez, Condesa, and Roma voice concerns over the project ("fucking

FIGURE 4.9. The closing moments of an anti-Corredor march along Avenida Chapultepec, near the Ángel de la Independencia on Paseo de la Reforma, December 5, 2015

hipsters" as the Simón Levy character calls them), "residents" from other neighborhoods are brought in to outvote them, and they are unceremoniously dismissed as the Mancera character praises the victory of democracy and citizen participation.[38] Just then, the *Supercívicos* fall through the ceiling of the set, smashing Simón Levy's beloved model and bringing his character to tears. They implore citizens to vote against the project on December 6, cutting to shots of the infamous second-story highway system on the west side of the city and explaining that this noisy landscape of cars and concrete will be the fate of their neighborhoods should the project be completed. They also make the important point that the Corredor was but "the first of ten projects that will convert our city into this [the 'second story' highway landscape]," placing further emphasis on this vote as a crucial first encounter in what would likely be a long and protracted city-wide struggle (*Los Supercívicos* 2015). This variety of organizing spaces and techniques yielded an interesting coalition. Using the traditional tools of movements like the occupation of space and the distribution and display of print materials, along with the virtual tools of contemporary social media (especially Twitter, Facebook, and YouTube), residents' opposition grew to include a diverse group of persons ranging across the spectrums of age, education, and any number of other social indicators, and to some degree spanning the city's varied geography. As these tactics indicate, the first battle to be won was that against the despondency of local residents, the second against the project itself.

For their part, Mancera, Levy, and the rest of the project's supporters conducted a similar campaign. Levy made numerous public appearances and gave several television interviews extolling the virtues of the project. ProCDMX placed large tunnel-like wooden structures containing beautiful visual renderings of the lineal park together with information about the project and its significance to the city in several public areas.[39] Bearing the colors of the city government, these archways appeared to be equal parts public information and project propaganda. Some were soon vandalized, their state-sponsored messages covered over with counterclaims (see Figure 4.8). Still, the Mancera/Levy marketing campaign was highly successful in certain respects, including especially branding. I found this out when I first began to ask questions about the project in the presence of opposition activists. I had to be told, more than once, that I was using the state's own language when I referred to the project as the triple-C (as the Zona Rosa portion of the name quickly disappeared from most advertisements and the project became simply the Corredor Cultural Chapultepec, or el CCC). In an attempt to respond to the state's effective marketing, residents generated several counter-brandings, including Shopultepec (a nifty anglicized neologism) and Centro Commercial Chapultepec (Chapultepec Shopping Mall). Parallel campaigns against the project used such slogans as "No CORRedor, No CORRupcion" (No corridor, no corruption) and "Así No" (Not like this).[40]

Mancera and Levy also had the powers of the state on their side, which had several implications in this case. It was of course a (nominally independent) state agency, the IEDF, that set the precise terms of the vote. Many project opponents claim that the date of the vote, for instance, was chosen in an effort to limit voter turnout, as many residents reportedly take out-of-town vacations over the weekends in early December. Similar claims of manipulation were made about the location of polling stations, which were said to be strategically located to make them difficult to reach for voters in areas where strong opposition to the project was expected. These locations were not formally announced until three days prior to the vote on December 3, when an interactive map appeared on the IEDF website. Tweets, retweets, Facebook posts, and other social and traditional media notifications spread across the digital networks of the city as residents sought to get out the vote. Anecdotally, rumors and accounts of IEDF website irregularities—including my own—circulated around the polling stations on the day of the vote. I had spent the previous day scrutinizing the IEDF map so that I might make an effective election-day canvas of polling stations. The morning of the vote, however, I was unable to access the map, receiving only an error message on the IEDF website. Local news sources also reported fraud allegations throughout the process, from the days leading up to the vote through the afternoon of the fateful day, including especially vote-buying.[41] The validity and extent of these claims, some of which can be reasonably substantiated with first-hand accounts, remain somewhat elusive. It is significant, however, that many accusations and arguments stem from or affirm belief in the PRD's ability to control the voting process and the clientelist machinery of the neighborhoods around the historic center in particular.

The most important and interesting of all the state's decisions pertained to the scale and geography of the vote, which was to be taken at the level of the whole of Delegacion Cuauhtémoc.[42] This choice made little sense to residents initially, as the delegacion comprises some sixty-five colonias, only three or four of which stood to be directly affected by the project, and only two of which technically bordered it. The LPC allows the consulta to be convened at nearly any conceivable geographical scale and configuration, leading many to wonder why sixty-five neighborhoods would be surveyed when only three to four stood to be directly affected. The official reasoning was that the project, in using some public monies and being pursued through a quasi-public agency, stood to affect residents well beyond the neighborhoods immediately bordering the avenue. However, this justification obscures the intentionality of this choice in two ways: the rationale of public accountability in no way implies the delegacion as the correct scale for the vote (if anything, this would seem to imply the Federal District as the optimal choice); and, less verifiably, Cuauhtémoc contains several neighborhoods in which the PRD enjoys particularly strong support, or is said to be especially able to corrupt the voting process by buying votes, intimidating voters, or engaging in other kinds of electoral manipulation characteristic of both PRI and PRD governments for decades and of the kind anecdotally reported by local sources on the day of the vote.[43] As I was told by a ProCDMX official, the ultimate result of the vote was received there as a highly unfortunate surprise, as Levy and Mancera had done all they could to plan for and produce a favorable, or at least negligible, result.[44]

Thus, the planning process surrounding the Corredor project is productively viewed as postpolitical in several senses. The project was developed by elites beyond the influence or knowledge of residents, legally mandated citizen participation was limited to ineffectual consultation about decisions that had already been made, and, as had overwhelmingly been the expectation among even the most active of opponents, even the miracle of a highly unfavorable result in the December 6 vote would likely have produced little more than a few condescending words about the value of citizen input from Mancera and Levy, who would nevertheless have gone ahead and erected their planned masterwork atop the avenue. The vote, however, did not go as the concerned authorities had planned. Rather, the final count was nearly two-to-one against the project (35.3 percent in favor, 63.5 percent against, 1.2 percent null), and by day's end Mancera had publicly announced (via Twitter) that his government would respect the vote. News sources immediately reported the verdict and the unexpected decision to treat the vote as legally binding, and several days of tension concerning the project ensued. Questions swirled as some residents celebrated and others cautioned that nothing was yet certain. Mancera's concession, not to mention the result of the vote itself, had been so unexpected that residents wondered whether it could be trusted. As it turned out, as I was told by a ProCDMX official some months later, the consequences for the planning process in Mexico City were far more pervasive than even the most hopeful supporters had dared to dream: Mancera immediately tabled the remainder of the rumored ten megaprojects.[45] Announcements soon circulated that the city would explore new options for the avenue, including promises for greater citizen input throughout the next planning process.

Even amidst the celebratory atmosphere that had begun to set in by the time of the victory gathering in Juárez one week later, however, uncertainty persisted about the shadowy machinations and dealings that had so unexpectedly produced this joyous result.[46] While residents and observers could call to hand a ready set of answers, no solid body of evidence could explain why Mancera had chosen to treat the vote as if it were binding, given how carefully the process had been managed to afford his government its freedom of judgement in the final analysis. A common line of thinking, and the most compelling, is that Mancera had long planned a national presidential run, and did not wish to alienate voters by flatly denying their will in what had become a very public contest. Disregarding the Corredor vote would have hurt him as a candidate, especially given the rhetoric Mancera has employed during his campaign for and tenure as jefe de gobierno. As with the establishment of the PRD's growing political control of the city through the 1990s, Mancera's path to the top executive post in the city was laid with the paving stones of continuing democratization. His official campaign slogan, "Decidamos Juntos" ("Let's decide together" or "Deciding together"), carried the image of a man of the people while adeptly sidestepping the neopopulist cult of personality forever attending the complex figure of AMLO. This image could also be read as an attempt to retain the leftist appeal of the PRD base in the face of the bitter disillusionment many residents express in their assessments of Ebrard, whose decision to ignore citizen pleas and the stern admonishments of the CDHDF in the building of the Supervía Poniente severely tarnished his reputation.[47] Possibly as a result of his aggressively democratic message and promises to remain democratically above the fray, Mancera has enjoyed a robust popular mandate since his landslide 2012 election, in which he garnered a share of the vote 43.85 percent larger than the next leading candidate, Beatriz Paredes Rangel of the PRI/PRVM coalition.[48] No mayoral candidate had yet achieved such overwhelming support, not even the mythic Cuauhtémoc Cárdenas, whose margin of victory was 23.5 percent in the city's first mayoral elections in 1997.[49] In order to continue his political career, this argument goes, Mancera had to rescue what remains of his image as a politician whose respect for democratic processes could be trusted. This explanation thus compellingly places Mancera's ultimate consulta decision within the carefully crafted narrative arc of his quite successful political career.

CONCLUSION

In framing Avenida Chapultepec as a problem requiring intervention, in developing the Corredor Cultural Chapultepec as the solution, and in its implementation strategy, the Mancera government and its allies utilized and relied upon their postpolitical hegemony in Mexico City. The mechanisms and apparatuses of this political order include the LPC, purported PRD control of the voting process, and many other legal and extralegal functions of state and party. Mancera's decision to honor the December 6 vote did not make political what had been not political, however, but only revealed the actions of the residents concerned as having always been the enactment of politics, that spectral imminence Rancière maintains is the omnipresent political potential of a radical remainder. Mancera's postpolitical order was not apolitical, therefore, but rather hegemonic. As the unfolding of the Corredor process and the ultimate abandonment of the project have illustrated, hegemonic relations are unstable and susceptible to ongoing assault by those forces bent on establishing a different hegemonic relation, a different police order with its own perceptual cartographies. The history of the PRD's break from the PRI, its efforts to gain and maintain control of Mexico City's executive and legislative powers, and its continual production and defense of legitimacy in the years leading up to the Corredor episode (as discussed here and in Chapter 2) betray a host of contentious activities only rightly understood as political, and as this political order's conditions of emergence. But rather than reject the participation process entirely, residents and their allies strategically exploited its weakest point, striking at the heart of the democratic legitimacy so prized by its figurehead. Even holding aside Mancera's political aspirations, the fact that his partially crafted, partially inherited postpolitical hegemony had been built upon foundations of democratic participation made it vulnerable to precisely this political gesture performed in precisely this moment and place. Rather than succumb to their own dismal appraisal of political realities, resisting citizens pushed what was largely considered a powerless vote beyond Mancera's breaking point, and succeeded in bringing down a decidedly postpolitical initiative from within the established rules. This victory suggests that resistance to depoliticizing orders in any variety of urban contexts should look for strategic vulnerabilities specific to their conditions of emergence. This enactment of grassroots politics also suggests a potential third response to failed institutions and oppressive postpolitical urban regimes more generally, a strategic alternative to the violence that can seem inevitable in the face of denials or repressions of equality (Dikeç 2017), or invitations to principled abstention from comprised spaces of dialogue that sometimes verge on sanctimonious.[50] Though undoubtedly an insight the utility of which is highly circumscribed by the particularities of chilangolandia, it nevertheless assertively suggests the relevance of spatio-temporal context (i.e., geography) across an urbanizing planet.

In order to exploit this democratic vulnerability, resisting citizens had first to combat the crippling despondency and cynicism palpably evident throughout the city, and thereby to recover a critical mass of revolutionary zeal. Voicing and enacting dissent and resistance were rights legally recognized and protected by the LPC. Convincing vecinos that these were rights worth exercising, rights with any real substance or power, however, was a difficult undertaking. By a variety of innovative means deployed across digital and material urban space, significant energy was focused on disrupting operative notions of citizen positions and the role of political participation. This suggests an attention to an underappreciated, but highly significant area for the study of postpolitical or depoliticizing regimes. As in the important related controversy over the city's planned airport explored by Davis and Rosan (2004) and Stolle-McAllister (2005), those opposed to the Corredor project refused the place assigned them, the role of anemic participants in what seemed a hollow democratic exercise. Their response, however, was not simply to abstain. Instead, they transformed what the GDF had intended as a limited dialogue within predetermined spaces and on prearranged terms into an active organizing campaign, expanding the scope of participation to include the extensive use of multiple media to express their messages and reimagine the strength of their collective voice. There was no physical violence nor threats of such violence, as had been the case in the airport saga, though the subjective violence of identity realignment was profoundly felt by residents in the aftermath of the vote. The following lines from a quiet, middle-aged woman who waited patiently to stand and deliver her perspective at a neighborhood meeting in Juárez some months later, accurately capture this shift in attitude (fieldnotes, April 24, 2016):

> I'm from Juárez. Before December 6, I didn't know so many residents from Juárez, Roma, or Condesa. Now, because of the actions of Simón Levy, we shout together, we struggle together, we know each other. Now, we are strong. Together, we are not less than Mancera, Peña Nieto, no one. Before December 6, we didn't know that.

Less than a month after the Corredor project was scrapped, the city that had been known as the Distrito Federal became overnight the Ciudad de México, as a host of sweeping changes brought on by a revolutionary wave of political reform began to take hold. Soon, much of the city's grassroots would turn its attentions to the drafting of its first political constitution (the subject of Chapter 5), playing the role of a revolutionary demos pushing through the seams of the social envisioned by the city's prophets of progress, and riding the high of the sunny December Sunday when they shouted down the tyranny of TINA by the sheer strength of their numbers.[51]

CHAPTER 5

The Redemptive (Urban) Revolution

Political Reform and the Rebirth of the Capital City-State

Ought we to affirm that democracy (like freedom, equality, peace, and contentment) has never been realizable, yet served (and could still serve?) as a crucial counter to an otherwise wholly dark view of collective human possibility? Or perhaps democracy, like liberation, could only ever materialize as protest and, especially today, ought to be formally demoted from a form of governance to a politics of resistance.
Wendy Brown (2011, 56)

Amidst the anaemiating fog of the record air pollution already accumulating in the early weeks of 2016, President Peña Nieto announced the consummation of constitutional reforms coveted and demanded for decades by the capital's residents and grassroots militancies. The following morning on Saturday, January 30, a photo of the president warmly embracing an enthusiastic Jefe Mancera dominated the front page of leading daily *La Jornada*, under the headline, "Reforma del DF, Triunfo del Pacto por México: Peña" (Reform of the DF, Triumph of the Pacto por México: Peña).[1] Among a host of other changes, Peña Nieto's 2016 reforms included extensive amendments to Article 122 of the 1917 constitution that dissolved the Federal District and created a new, thirty-second federative entity known as la Ciudad de México (Mexico City) within the Mexican Union. This stroke of the pen not only rechristened and reclassified the capital city, but also undermined the prized exceptionalism of many a *defeño*. Alongside this topophilic affront, and owing only in part to what many everyday chilangos experienced as a vague presentation and uneven rollout, the reforms seemed to conjure stubbornly immiscible affective impulses. These impulses often translated into bellicose appraisals that uneasily suspended the sobriety of longstanding distrust alongside the strident hope of a seemingly indefatigable remnant of the grassroots reinvigorated by the flickers of latent political energy they had begun to see among their neighbors in the wake of the Corredor saga. Most of my friends, informants, and colleagues seemed to think that it was an exciting, if confusing, time to be a chilango.

This chapter delves deeper into the quotidian significance of the suite of changes commonly and collectively known as political reform. Introduced in Chapter 2 as a major part of the capital's democratization in the midst of a growing tension between the capital city and the national state under the PRI, political reform provides a productive lens through which to view the contemporary manifestation of the long-simmering tensions between scales of authority and bodies politic—specifically the urban and the national—the last of the trajectories of conflict also introduced at the conclusion of Part I. I begin this examination with four ethnographic scenes which, taken together, illustrate some of the contours of the charged blend of excitement and trepidation palpable in the city prior to the election of the Constitutional Assembly tasked with giving form and life to the city's first constitution. Elements and shades of unease seemed to stem in large part from either a general distrust in a duplicitous Mexican democracy or from a more specific fear of the PRI and the spectral dangers of its imagined political schemes. Persistent enthusiasm, by contrast, seemed to spring from a variety of sources; from a reinvigorated sense of the progress and promise of democratization to the obstinate faith that even if the worst visions of PRI treachery were true and the constitutional process really was nothing more than an elaborate sham, it still presented the city's residents with a potentially historic opportunity. That is, even if PRI (and/or PRD) leadership intended to use the constitutional process for sinister political ends, the democratic moment its birthing would necessitate was a juncture of uncertainty similar in nature to that of the Corredor project but much grander in scale. This was a reality some residents and civil society leaders stood ready to exploit.

To make sense of the tempestuous mixture of plans, promises, and prophecies uneasily circling these events in 2016 and 2017 (and the feelings that fed them), I will follow the discursive, political, and temporal linkages made to the Mexican Revolution and to notions of revolutionary politics more generally that were circulating with impressive frequency around Mexico City in 2016. In order to parse these ties and the uses to which [R]evolution seems to forever be put in the capital city's politics, I will consider political reform from its ideational inception at the local level and the national reforms of the late 1980s and early 1990s to the adoption of Mexico City's first constitution by the Constitutional Assembly in early 2017. The notion of political reform in the capital *as* revolution will propel this investigation through several dimensions to be explored in the sections that follow. In the third section, I will trace the unfolding of the final stages of political reform in the capital, building on the tensions elaborated in previous chapters. The fourth section will consist of a multivalent analysis of these processes through the lens of revolution and the multiple refractions it provides for the parties involved, from residents of the city's peripheries to party bosses and real estate moguls. This analysis, focused on what I collectively refer to as a revolutionary structure of feeling (following Raymond Williams), will provide a useful approach for understanding the seemingly contradictory mixture of thoughts and feelings surrounding the process of political reform in the run-up to the delivery of the 2017 constitution and the momentous elections that were to come in 2018. I will build on this analysis to argue that many of the lines cut through this structure by various guises of revolution are aimed in some way at a redemption of the tragic promise of revolutions past, along the messianic lines proposed by Benjamin ([1968] 2007). In the concluding section I argue that in this sense, whether their efforts sang with the sharpest sincerity or smacked of unctuous Janus-facery, they shared an interest in making this urban revolution the truest Mexican Revolution of all.

POLITICAL REFORM AND EVERYDAY LIFE: FOUR SCENES FROM A STRUCTURE OF FEELING

Nearly four-and-a-half months after the small December celebration in a sleepy corner of Juárez where residents and their allies met to celebrate the victory of their campaign against the Corredor Cultural Chapultepec, another demonstration took place in Mexico City, this time with more obviously national origins and implications.[2] On the sunny Sunday of April 24, 2016, a massive crowd took over the Paseo de la Reforma in a call for greater support and attention to the value and the rights of women in Mexico. Many adorned with shirts bearing the names of organizations, parties, or lost loved ones, and some bearing no shirts at all, an inestimable number of women marched from the Monument to the Revolution to the Angel of Independence. Some sang, others laughed, and a few wept. I count myself fortunate to have witnessed some of their march. I had planned to attend

FIGURE 5.1. Women's march along Paseo de la Reforma, April 2016, with the base of the Ángel de la Independencia visible in the background. Photography by author

the event with my friend Marí, an incisive critic and feminist of inspiring praxis, though we never found each other in the crowd (fieldnotes, April 24, 2016, and see Figures 5.1 and 5.2):

> The Angel had already begun to crowd with people when I arrived. . . . The eastern side was filling up quickly with woman of all ages wearing purple and white t-shirts bearing a variety of feminist messages. Many carried flags or signs. There were also some men, some children, and some reporters and lookers-on. I found a good spot at the height of the steps, dead center. I took a few photos as I waited. The crowd was interesting, and there were some justifiably intense statements being made to reporters here and there, about the widespread violence against women in the country, about the lack of justice for women, about women being equal political voices, etc. Many of these statements were being made by young women, I found, but there were plenty of *abuelas* in the throngs as well.

Soon, the marchers approached, overwhelming the westbound side of Reforma while lines of (mostly female, interestingly) traffic police tried desperately to hold them to that side. Traffic continued to flow along the eastbound side. Soon enough, though, the march would invade that side as well and traffic would be halted entirely. The crowds began to fill the glorieta, to wild applause and a multitude of cheers and chants. Some women walked

FIGURE 5.2. Women's march, April 2016, looking roughly northeast along Paseo de la Reforma. Photography by author.

into the traffic massed at the south side of the glorieta, wading with signs and chants into the captive autos, pointing and shouting their message. Police whistles only served to heighten the general din.

. . . I came down around the south side of the Angel's perch and picked my way into the street. It was a comfortable chaos. I immediately noticed a great variety of women's groups and diverse agendas. . . . There were some side-long glances of angst and suspicion as [groups] bumped against each other at the pinch point of the march, where *Reforma* meets the glorieta. The take-over of the eastbound lanes had helped alleviate some of this pressure, as the marchers spread out across the boulevard, but still the tension found expression. Even so, a mood of acceptance and collegiality pervaded the event, and I never witnessed anything more than glances and murmurings among the participants.

For the men, it was another story. Though most groups had stated that all sympathizers were welcome, there were some groups for whom there were far too many men present. At some point, when the congestion had reached a head, a violent chant and dance broke out. A line of photographers had formed, mostly men, at precisely the wrong point at the glorieta/ Reforma junction, and was seemingly contributing to the slowing of the march, particularly as it was difficult to get large signs through them. Frustrated by this, a group of young women formed a cluster that faced inward and began to aggressively chant "Solo mujeres!" As they chanted, the surrounding masses began to take notice. The chant grew in strength, and the group began to push through the congestion, forcing photographers and everyone else aside as they, at long last, punctured the glorieta. Wild applause followed, and a prideful euphoria seized the terrain.

One of the many groups I spoke with at this event wore t-shirts emblazoned with the logos and slogan of AMLO's Morena, "La Esperanza de México" (The hope of Mexico). They told me that they belonged to the leadership of the women's arm of the party. Their shirts, and those of several other groups, however, also carried another message, that of the importance of the upcoming election of members of the Constitutional Assembly, a group of one hundred persons who would decide the fate of the city's forthcoming constitution. Others, too, wore t-shirts with constitutional messages, the most common of which stated, "For a democratic, popular, citizen, and FEMINIST constitution!" As I spoke with marchers and onlookers, it became clear that the city's constitution was viewed by some hopeful residents as a significant opportunity for meaningful progress on women's issues, among other pressing concerns.

As the march was beginning to clear in the early evening, I walked back into the Zona Rosa and took the Metro one stop across Juárez to attend a meeting of what its organizers called the Juárez Neighborhood School of Citizenship. These meetings were held in a stately old building on Turín street then still adorned with vestiges of the anti-Corredor campaign waged the previous fall. In a crowded room on the second floor, neighbors met on select Sunday evenings to debate the politics of their city and country, to ask questions of one another and hear the insider perspectives of those attendees most active in local politics, and to be educated on such topics as the history of rights and modern citizenship in the western world, as was the initial topic on this particular evening. Freshly printed photocopies of shabby versions of the French "Rights of Man and Citizen" and the US Declaration of Independence were distributed as the two women leading the discussion extolled the virtues of a historical perspective on rights and the significance of Mexico City's current political moment. Even their best efforts at promoting the institutions of democracy as useful channels of power available to citizens, however, had trouble penetrating the fog of skepticism and distrust that seemed to permeate every discussion. Recent daily coverage of the so-called Panama Papers had only fueled this mood, and even one of the organizers at one point categorically stated that "Mexico is a paradise for the corrupt" (fieldnotes, April 24, 2016). Still, the organizers fought for the idea that legal documents and philosophical antecedents like these declarations, along with international treaties and local and national law, have great relevance, at least in that "governments sign these documents, and then they are obligated to follow them, and to make laws that support them!" (fieldnotes, April 24, 2016):

> a [man] seated in the opposite corner of the room bellowed his take on this issue of accountability for politicians: "Corruption is all powerful, and total, throughout this country!" This began a series of exchanges between him and the two women behind the table. At one point, [one of the women] began explaining, "In the United States, it works like this . . ." As she continued, [the man] interrupted her: "Yes, but we are not the gringos!" This was received with wild laughter and some light applause, and a few sad looks. It cut the woman's idea down in its tracks, and the conversation moved on.
>
> [The other woman] behind the table pushed the meeting to recognize the importance of collective strength. "Now we're organized," she stated confidently. "We're not organized!" retorted [the man]. With a look of irritation, she responded "We're organizing right now!" "We *want* to be organized," he stressed. After a thoughtful pause, the woman specified "We are *in the process* of organizing." Silently, but with a slight bow of acquiescence, [he] relented.

Here too, the conversation turned eventually to the city's constitution, with Herón and the other organizers imploring their neighbors to pay careful attention to the process. Some of the most significant actors involved, Herón explained, were beholden to or even directly represented commercial or political interests (implying a certain hostility to the interests of local residents). This allegation resonated with most of the group, it appeared, and fed what felt like the beating of the dead horse of corruption. As with the Corredor project, however, the process of establishing the Assembly required a democratic moment, in that some of its members would be voted on by the city's residents. This presented residents, Herón argued, with an opportunity to force some semblance of accountability, and perhaps to make a meaningful contribution to the constitutional process. At this meeting, the intoxication of the victory over the Corredor project seemed to have worn off somewhat, despite the persistence of a victorious rhetoric, and mingled uncomfortably with general skepticism. Despite the organizers' efforts, talk of corruption seemed to animate these residents far more than talk of participation in the constitutional process, the energy of critique and cynicism finding a strength then-unmatched by that of creativity and hope.

The following day, I attended a meeting at El Museo de la Ciudad de México in which several panels were convened to discuss the constitutional process (see Figures 5.3–5.6).[3] The panelists were largely academics, activists, and civil society leaders, and the evening's keynote address was given by Cuauhtémoc Cárdenas. The last panelist of the long day of speakers was Sergio González, an architect by training and a politically active resident of Colonia Juárez. Frankly stating the concerns of many residents, he forcefully argued at the outset:

> This constitutional exercise is not anything democratic, for at least two reasons. On the one hand, the elucidation, debate and decision on the laws that we require is being carried out in haste, as the priority of the promotors is to synchronize such exercises to sexennial schedules and political objectives that are far outside the schedules of citizens. We are not in a hurry to get a Constitutional Assembly in six or eight months. And we are prepared to take all the time that is necessary for a project of such magnitude. On the other hand, it is an exercise without the direct participation of the socioeconomic and cultural diversity that characterizes the inhabitants of this city. The timetable of the political class and selective and vertical participation express a conception of democracy reduced to procedures and demographic polls: surveys, virtual consultations, opinion polls, forums like this one, which obviously excludes majorities from participation in the elucidation, deliberation, and decision-making of laws.

Other panelists struck a more hopeful tone for the process. Making reference to the "Así No" slogan of anti-Corredor forces, one panelist insisted that as valuable as such explicitly critical and oppositional discourse could be, there was also a necessity for an "asi sí," a force to imagine and to defend these neighborhoods, and by extension the city, both as they were and as they could be. He and others argued that the coming constitution provided a rare opportunity to exercise such an affirmative voice. Voicing agreement on this point, Marco Rascón Córdova, a founding member of the PRD, provocatively asserted that the constitution should be "not an end, but a beginning" (fieldnotes, April 25, 2016). For his part, Cárdenas espoused a measured optimism with regard to the constitution's potential, but likewise cautioned that though the constitution promised to regulate many issues in general terms, much of the real political work would take place in the

FIGURE 5.3. Marco Rascón speaks alongside Dulce Colín at El Museo de la Ciudad de México. Photography by author. **FIGURE 5.4.** Sergio González speaks at El Museo de la Ciudad de México. Photography by author.

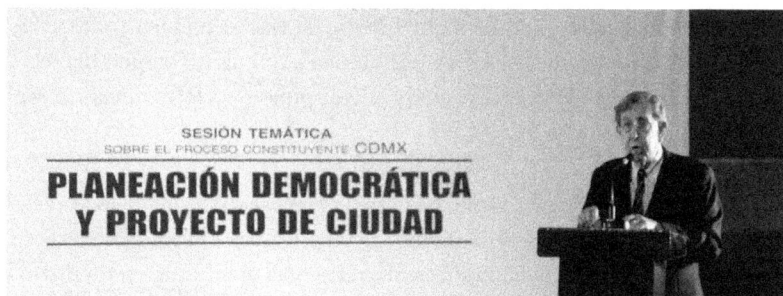

FIGURE 5.5. Cuauhtémoc Cárdenas delivers a keynote address at El Museo de la Ciudad de México. Photography by author.

FIGURE 5.6. An audience member speaks at El Museo de la Ciudad de México. Photography by author.

secondary or supporting legislation drafted in its wake. His tone was less enthusiastic than many seemed to have hoped for, and his wide-ranging critiques of the process and cautionary notes about the state of urban development and urban governance in the city seemed to leave many in the audience with a somewhat negative, if realistic, impression.

Some days later I found myself in an Uber on my way to an interview, trying to make sense of this range of opinions and to focus in on the political story of the constitutional process. I had been repeatedly told by friends and strangers alike that the whole process was a trick of the PRI, an elaborate scheme developed by party elites aimed at the reactivation or reconstitution of the party's machinery in and control over the Federal District.[4] In the midst of a casual iteration of a near-daily conversation in Ubers or taxis about my purpose for being in Mexico and my feelings about the capital city, I decided to broach the subject with the driver. Like many other older Uber drivers I had met in Mexico City, he was or had been employed in the professional ranks. Unlike many others, he was in no hurry, and even indicated that he wanted to finish our conversation as he tapped his telephonic completion of the trip's charges and pulled up to my destination (fieldnotes, May 3, 2016):

> He told me that as the DF would be treated more like a state in the wake of the reforms, it would lose some of the special privileges it held as a district, including some of its autonomy from the remainder of the country. He also had a lot to say about the PRI and its supposed return to the city through the reforms. "Of course it's a trick!" he said, in response to my saying that I had been told that the constitutional process "is a trick, of the PRI, in order to reconquer the Federal District." That he firmly believed the process of "political reform" to be a "trick" in this way, however, was no obstacle to his thinking it a significant moment. It may be a trick, he reasoned, but it was still an opportunity to remake the city in spite of this. The PRI had lost the city decades ago, he explained (in a familiar refrain), and desperately wanted to get it back, as they had the national government with the election of Peña Nieto. The problem, he argued, was that *chilango* support was less easy to purchase or control than that of the rest of the country. "In the rest of the country, you can easily buy votes. But here in DF, people will take the money, but then vote how they want anyway."

Irrespective of this person's appraisal of the process of vote-buying in the city vis-à-vis the countryside, his perspective is clear in its illustration of the profound coexistence of seemingly contradictory attitudes expressed in each of the other three scenes related above with regard to the constitutional processes of 2016 and 2017. Treating the process as a trick, a ruse, may have felt necessary in order maintain critical distance and, perhaps, to stave off another heartbreak of postpolitical democracy, akin to the relief derived from the flexible and ambiguous power of humor.[5] But to completely dismiss the constitutional process was equally dangerous, many argued. To leave this ruse to run its course would mean the surrender of the capital city and its neighborhoods to ill-intentioned forces of monstrosity, from a resurgent ruling party to the real estate cartels. It would be to allow their homes and communities to be carved-up and horse-traded by those they considered the wicked masters of graft, to turn the other cheek as Mexico City became a gilded hellscape openly and uncontestedly derided as a paradise for the corrupt.

In conversations and interviews, at public events and in private meetings, in press accounts and even in Twitter beefs and combative YouTube comments sections, I often found that, if pressed, even the most obstinate would acknowledge the complexity of this conundrum. Political reform and the constitution it promised were not simply a trick, nor simply a tool, a prize, or a golden opportunity. Rather, in the spring of 2016, political reform represented each and all of these possibilities. The constitutional process was therefore an open terrain, a privileged site for the staging of a politics that would surely have reverberations for untold decades. Building on the historical geography of political reform presented in Chapter 2, the following section details the incremental and episodic process by which the culminating chapter of political reform became a reality in Mexico City in 2016 and early 2017.

LA CIUDAD DE MÉXICO: POLITICAL REFORM AND URBAN REVOLUTION

At the outset of his tenure as jefe de gobierno, Miguel Ángel Mancera's participation in Peña Nieto's Pacto por México was most likely born of several complementary motivations. Partnership with a resurgent PRI was not a popular move among the city's grassroots, but did serve to bolster a party already floundering nationally and increasingly feeling the pressure locally from the fledgling movement that would become AMLO's powerful Morena (to say nothing of the more prosaic suspicions and rumors related to Mancera's own national aspirations). What has frequently been overlooked, however, is that Mancera's interest in the Pacto was explicitly framed as the culmination of the incremental political reforms of past PRI presidencies. Previous mayors, including AMLO and especially Ebrard, had been making noise about the city's full emancipation from the federal state for more than a decade by then, as discussed in Chapter 2.[6] But what Mancera asked of the president was the full breadth of changes first envisioned by the city's residents and progressive leaders in the late 1980s, full political autonomy and an equal place in the federation as the union's thirty-second state. Just over three years later, Peña Nieto largely delivered on this promise by signing the reforms of January 2016.

The changes to Article 122 of the Constitution of 1917 were far-reaching, abolishing the Distrito Federal and creating a new, autonomous *entidad federativa* (federal or federative entity) within the Mexican Union to be known as la Ciudad de México. The language used to create this new entity caused a great deal of confusion, specifically around the issue of statehood. Officially designating the capital a federative entity allowed Peña Nieto to avoid the political quagmire of officially creating a thirty-second state, but immediately raised questions about the precise nature of the new territory's autonomy, its rights and duties to the Union, and its position vis-à-vis the other states.[7] Months of debate by legal scholars and contradictory language and reporting by local media (not to mention activists and local political authorities) hardly promoted a healthy general understanding. Nevertheless, and despite a fog of public perplexity, the city had been granted considerable political autonomy, and in practice became a state in nearly all but title.[8] Much of the city's governmental infrastructure remained intact, though with some important titular and legal alterations. The sixteen delegaciones were converted into *alcaldías* (typically translated as "mayoralties"), rather than municipios, and their executive bodies were also granted more political autonomy from the city government. The reforms therefore accomplished a considerable decentralization of authority, a process begun in the 1980s, despite stopping just short of elevating the capital to formal statehood. Many of the thorniest questions about how the new city would be governed and relate to both the federal state and the new alcaldías, however, were left to those tasked with crafting the city's first constitution.

In addition to requiring that a constitution be proffered by the sitting jefe, Peña Nieto's reforms created a series of deadlines and a set of procedures for the drafting and formal adoption of this document. The constitution was to be debated and eventually adopted by an assembly of one hundred persons, which was to convene in September of 2016. Of these, sixty *constituyentes* were to be directly elected by the former Federal District's residents on June 5, 2016 (via proportional representation), and the other forty were to be appointed: fourteen by the Federal Senate, fourteen by the Federal Chamber of Deputies, six by the mayor, and six by the president.[9] This Constitutional Assembly, however, was not tasked with initially creating the constitution. Keeping with the tendency to concentrate power in the executive, the reforms allowed only the jefe de gobierno to present a draft constitution to this assembly, to which they could then make minor changes. To accomplish the task of creating a draft constitution, Mancera convened a group of twenty-eight *notables* (notables), which included politicians, legal scholars, university professors, activists, and architects, which was announced within a week of the reforms.[10]

The drafting process also had a highly touted citizen participation component, which was promoted through public meetings and several online platforms. At innumerable public events held throughout 2016 (but especially in the spring months, before the June 5 election of constituyentes), everyday citizens were encouraged to ask questions about the constitutional process, make suggestions for the city's constitution, and hear from elected representatives, civil society leaders, and other proponents and opponents of the process from all political angles. Some of these gatherings were put together with the support or in the name of the city government, or one or more political parties, but often they were organized by independent groups and institutions, including universities and research institutes, NGOs ranging from local to international in scope, and neighborhood associations.[11] I attended several such meetings, including the public launch of the city's *Espacios de Encuentro* (Spaces of Encounter) strategy, which featured notables Alejandro Encinas Rodríguez and Porfirio Muñoz Ledo,[12] among other speakers (see Figures 5.7 and 5.8). The bulk of the crowd at this event was made up of a small army of uniformed youth volunteers who would be charged with monitoring and offering assistance to those seeking to use the individual stations (Points of Encounter) to be set up at some three hundred locations all around the city and freely available for citizen use. These stations consisted of glossy cardboard podiums containing iPads that would lead citizen users to the online platform through which they could craft proposals to be considered by the notables and the Constitutional Assembly, ask questions or offer comments on the proposals of others, and learn more about the constitutional process (see Figure 5.8). Though much of this type of participation was to be funneled through Change.org, which could theoretically be accessed from any device connected to the internet, the Points of Encounter strategy was aimed at facilitating the participation of those residents without private access to this platform, whom the volunteer monitors would be able to assist should they need help navigating the technology.[13] At the height of this push for participation during the spring and summer months, however, it was often left unclear how this citizen input was to be collected and considered. I witnessed several encounters in which

FIGURE 5.7. A protestor interrupts a panel at El Museo de la Ciudad de México during the launch of the city's Spaces of Encounter strategy; March; 2016. The message reads; closed formal spaces + web page = simulation. Photography by author. **FIGURE 5.8.** Points of Encounter stations on display at El Museo de la Ciudad de México. Photography by author.

volunteers or authorities were unable to adequately answer resident concerns about this process, and cynicism regarding the efficacy of this initiative was rampant.[14] A common refrain among critics consisted of an attack on what the constituyente offered as participation, exemplified by a protestor who marched in front of the stage during one of the formal speeches at the Spaces of Encounter launch, holding a sign they hurriedly unfolded as they hustled to the front of the room, which read simply "CLOSED FORMAL SPACES + WEB PAGE = SIMULATION" (see Figure 5.7). This person sought to convey in their thirty seconds of protest what many had been saying for months, that the carefully scripted interaction made available to them was better called political theatre than authentic participation. Still, Change. org officially registered 348 petitions, the top ten of which ranged in support from over ten thousand to nearly fifty thousand as of July 2016 (*Dfensor* 2016), and many residents were reportedly pleasantly surprised to see their concerns ultimately represented. This level of citizen input offered considerable legitimacy to the process and brought a good deal of international attention.[15] The heralding of this progressive effort and grand experiment in democratic crowdsourcing was arguably far more valuable to both the beleaguered Peña Nieto and Mancera administrations than was the content of citizen's concerns—whether or not these ultimately made any lasting constitutional impressions—a calculus lost neither on analysts nor lay observers.

As it had in the Corredor process, political geography played several vital roles in the processes of political reform, in its guises as the *locus, modus,* and *finis* of political struggle. At the heart of the complex dance of Mexican politics over the course of the last several decades is a developing conflict between political parties, and even more fundamentally, between their associated territories of authority. While the PRI began to lose control of the capital city in the 1980s, it nevertheless fiercely coveted this geography and took measures to ensure continuing relevance there. Though it has not always been read this way, the set of concessions to democracy made especially in the wake of the 1988 elections and the loss of the Cárdenista faction of the party to the PRD certainly support the thesis that PRI leadership, however quietly, desired to retain an electoral presence in Mexico City and, perhaps even more to the point, to make a play for the city's electorate and offices at some later date if and when the opportunity should arise. Even before the 2016 reforms, many of those observers who viewed the constituyente critically claimed to hear the whispers of unspoken collusion and to see the fingerprints of party operators on the Pacto por México. The PRI, with the help of an unusually friendly PRD mayor in Mancera, seemed to be making precisely such a play for the city. The geography of the Assembly's appointment process (forty out of a hundred members being appointed), as many pointed out, slanted heavily toward the national legislative and executive powers, where the PRI was strongest (and where it enjoyed majorities in both chambers). PRI interests were also thought to have been given some measure of influence among Mancera's group of notables, though this is difficult to precisely qualify.[16] Peña Nieto and the PRI thus were afforded, both directly through the president's reforms and possibly indirectly by Mancera, a greater measure of influence on the constitutional process than many observers claimed they deserved or could have won through almost any process of direct democracy based exclusively at the scale of the former Federal District. As I suggested above, Mancera and the PRD were likely also motivated by territorial concerns. AMLO's Morena, often openly antagonistic toward the PRD and its leadership, was already rapidly gobbling up the PRD's electoral base in the city, undermining the PRD's ability to govern and threatening Mancera's political future. In this geographical squeeze, Mancera appeared to have chosen an unlikely partner in Peña Nieto and his PRI in order to stave off the onslaught of AMLO and Morena. This alliance, begun in the Pacto por México and further solidified with the 2016 political

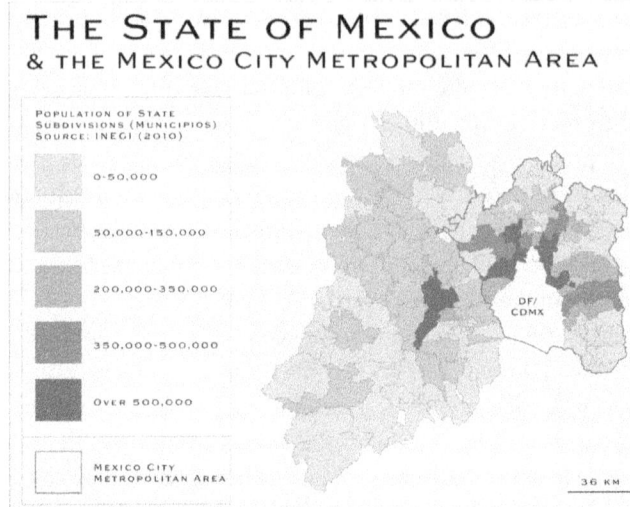

THE STATE OF MEXICO
& THE MEXICO CITY METROPOLITAN AREA

POPULATION OF STATE
SUBDIVISIONS (MUNICIPIOS)
SOURCE: INEGI (2010)

0-50,000

50,000-150,000

200,000-350,000

350,000-500,000

OVER 500,000

MEXICO CITY
METROPOLITAN AREA

DF/
CDMX

36 KM

FIGURE 5.9. Population density in the state of Mexico. Cartography by author.

reform, both promised Mancera a greater degree of control of the city in the face of Morena's challenge and kept his presumed plans for a presidential run at least theoretically viable.[17]

Perhaps even more interesting is the geography that was left unaltered by political reform's revolutionary upheavals. Though a generation and more of scholarship, observation, and advocacy has made clear that the city's explosive growth in the latter half of the twentieth century left the borders of the Federal District woefully inadequate as indicators of the functional extent of Mexico City and even more so as jurisdictional demarcations worthy of the city's many expanding crises, Peña Nieto's 2016 political reforms left these borders unchanged as the city transitioned from Federal District to *entidad federativa*.[18] Many had argued that these reforms provided the best conceivable opportunity to redraw these boundaries and consolidate the Valley of Mexico's various authorities into one political entity. The surrounding states of Hidalgo and México, however, would likely have fought such a proposal with the sternest vigilance, especially México, some 76 percent of whose population resides in the colonias populares, small villages, and other peri- and ex-urban areas of the metropolitan region and whose electoral politics formed a large portion of the beating heart of the PRI's national power even during the PAN sexenios.[19] The political cost of altering this geography was therefore simply too high, and the new city inherited its boundaries, as it did much else, unchanged.[20] As noted in Chapter 1, Rodríguez Kuri (2010) has shown that proposals for metropolitan expansion of the Distrito Federal, of precisely this sort, have been vigorously debated since at least the Constituyente of 1916–1917, wherein Carranza's proposal faltered on similar political ground (relations with the Estado de México). It is striking that the two moments in which this radical metropolitan ideal has come closest to realization should be spaced precisely a century apart, that both should be provisioned by momentous national constitutional debates in which the capital's transformation figured prominently, and that despite fundamental differences in both proponent's motivations (political domination in the post-Revolutionary context and political emancipation and metropolitan coordination in the contemporary context) and socio-spatial and political conditions (owing in large part to a century of rapid urban development that dramatically transformed the geographies of the valley, including, of course, the Estado de México) they should be defeated so soundly, so matter-of-factly, and according to such similar interscalar geopolitical logics.

Not unlike the geographical considerations of the broader political reform, the timing of the constituyente had far more to do with the perceived necessities and cruel vicissitudes of sexennial politics than with any rationale pertaining to either the requirements for assiduously crafting a robust legal foundation or the effective management of an ambitious program of citizen participation. As Sergio González argued in the public criticism cited in the introduction to this chapter, the process felt rather hurried. The whole saga, from the national passage of the political reforms to the Constitutional Assembly's acceptance of the finalized constitution, took exactly a year and a day, from January 30, 2016 to January 31, 2017. Indeed, the entire process seemed to pass a good many ordinary citizens by without much notice. Certainly, this can be partly attributed to the oft-noted and aforementioned apathy and cynicism regarding the city and country's politics, but the speedy timeline undoubtedly (and, for some observers, intentionally) played a significant role as well. Many residents I spoke with claimed to have only first heard about political reform when they saw advertisements posted on the walls of Metro trains and stations, part of the city's campaign to advertise its legal name change in the early months of 2016. These ads featured impressive photos of various city landmarks overlain with a simple message: "Adiós DF, Hola CDMX" (Goodbye DF, hello CDMX) (see Figure 5.10). In the months leading up to the June 5 vote, many of the city's billboards, whitewashed cement walls, and online platforms like Twitter and Facebook also displayed the names and messages of a variety of partisan voices vying for position in the Assembly. That many of these hastily installed promotional materials remained physically in place the following year, long after the process had concluded, speaks to both the frenzied

FIGURE 5.10. An advertisement on the Mexico City Metro, which displays the message "Goodbye DF, Hello CDMX." Photography by author.

pace of the constitutional process and to the somewhat artificial character of its timing. A low turnout (just under 29 percent of the city's electorate) at the June 5 vote, despite what seemed a serious push on the part of the major parties in particular, further contributes to this criticism.[21] By the fall of 2016 (the Assembly was formally seated in September), a draft constitution had reached the Constituyente, where its finer points were debated over the next several months. By the end of the year, press outlets had begun to criticize the seeming lack of progress the Assembly had made, and by the first weeks of 2017 faith in the Assembly's ability to adopt the constitution by the looming January 31 deadline expressed in Peña Nieto's reforms the previous year was a rare commodity. Much to the surprise of many in the city, Mancera's publicly expressed trust in the Assembly was validated with the approval of the new constitution at 1:47 a.m. on the final allowable day (Suárez 2017). After an uncomfortable nascency period that witnessed numerous legal challenges and the momentous 2018 local and national elections, the constitution entered into force in September of

2018, in the eye of the political storm walled by AMLO's July election and December installation.[22]

Published several days later after its acceptance by the Assembly in the city's official gazette and distributed in booklet form in the ensuing weeks of February, 2017, the city's first political constitution is unquestionably an ambitious and commodious work. Its brief preamble is introduced by the immortal words of Aztec ruler Tenoch, from 1325: "En tanto que dure el mundo, no acabará, no perecerá la fama, la gloria de México Tenochtitlan" (As long as the world endures, the fame, the glory of Mexico Tenochtitlan shall not end, shall not perish). The preamble goes on to assert that "the city belongs to its citizens," and unambiguously grants the constitution the mantle and accumulated capacities of a heroic political lineage, as "the culmination of a political transition of plural and democratic inspiration" and "historic resistance against oppression." The bulk of the remainder is composed of seventy-one articles and thirty-nine additional transitory articles which cover a wide array of topics in varying amounts of detail. The second chapter, consisting of Articles 4 through 14, is organized as the city's Charter of Rights. Beginning with Article 6, the document assures citizen rights by guaranteeing a city of certain qualities:

1. a "city of rights and liberties" (Article 6)
2. a "democratic city" (Article 7)
3. a "city of knowledge and education" (Article 8)
4. a "city of solidarity" (Article 9)
5. a "productive city" (Article 10)
6. an "inclusive city" (Article 11)
7. a "habitable city" (Article 13)
8. a "secure city" (Article 14)

Many of these same phrases appeared in the 2010 Mexico City Charter for the Right to the City, as the six strategic foundations that compose the CPCCMDC's vision of the right to the city.[23] In the city's constitution, the right to the city is ensured by its own dedicated article, Article 12. One of the constitution's briefest articles, its first section states, "Mexico City guarantees the right to the city which consists of the use and full and equitable utilization of the city, founded on principles of social justice, democracy, participation, equality, sustainability, with respect for cultural diversity, nature, and environment." Several persons familiar with the process explained to me that both the presence of the right to the city in the constitution and the obvious influence of the Mexico City Charter for the Right to the City's strategic

foundations on the constitution's Charter of Rights came as a direct result of
the participation of Enrique Ortiz Flores in the drafting of the latter text.[24]
That the concept ultimately found a place in the constitution at all, and particularly such a prominent position, should be considered a significant victory for those CPCCMDC members and other right to the city advocates
for whom its constitutional enshrinement had always been a primary goal.
Given its brevity and the vague and expansive nature of its language—to
say nothing of its somewhat odd placement among the other nine articles
that make up the Charter of Rights, which are each devoted either to particular sets of rights organized around the themes noted above or to speci-
fying the provision of these rights by the city government—Article 12 initially appears redundant and unnecessary. After all, the guarantees it seeks
to assure are all covered in far greater detail by other dedicated articles that
appear both before and after, some of which span several pages of text in
their elaborations. The second part of the article, however, arguably makes
a more significant statement: "The right to the city is a collective right that
guarantees the full exercise of human rights, the social function of the city,
its democratic management, and ensures territorial justice, social inclusion, and the equitable distribution of public goods with the participation
of the citizenry." Building on the twin pillars of complexity and collectivity
that guided the production of the 2010 Charter, the constitution here distinguishes the right to the city from the other rights it establishes by virtue of its pertaining not to an individual, atomistic citizen but rather to the
citizenry as a collective.

The Charter of Rights has received most of the praise offered by international observers, owing to the progressive entitlements, privileges, and protections it contains. Article 6 has received attention for its instantiation of
"the right to sexuality," which includes "the right to decisions about it and
whom to share it with," and freedom from discrimination based on "sexual preference, sexual orientation, gender identity, [or] gender expression
and sexual characteristics." Reproductive rights are also explicitly protected,
including "the right to decide in a free, voluntary, and informed manner
whether or not to have children." Articles 7 and 8, in addition to dealing
with many traditional rights of political expression and practice, make several interventions in the citizenship rights of the digital age. "The right to
information" detailed in Article 7 subsection D, which covers citizen access
to public information, is immediately followed by "the right to privacy and
the protection of personal data" in subsection E. Subsection C.3 of Article 8 asserts "the right to science and to technological innovation," which
includes "free access in a progressive manner to the internet in all public

spaces, public schools, government buildings and cultural sites." Other sections of the constitution offer definitions and principles for such diverse
areas of governance as the identification of historic sites and cultural heritage (including requiring an official oral history registry) (Article 18 subsection C), "the rights of original towns [or peoples] and neighborhoods and
resident indigenous communities" (Article 59), and combatting corruption
in the city (Articles 61–63). Article 25, which deals with the city's principles
concerning "direct democracy," clarifies and incorporates some aspects of
the Law of Citizen Participation (discussed above in Chapter 4), organizing
its organs of popular participation into five mechanisms: the "citizen initiative," the plebiscite, the referendum, the citizen (or "popular") consultation,
and the newly developed "revocation of mandate." This last apparatus allows
for elected officials of Mexico City to be removed from office "when this is
demanded by at least ten percent of the persons registered in the nominal
list of the electorate of the respective [geographical area]." The procedure
for the revocation of mandate, the citizen consultation, may only be made
once for a given official, "when at least half of the duration of the [term] of
popular representation in question has passed." The opportunity to remove
elected officials at such a low threshold of public incertitude has produced
no small measure of heated discussion within the city and beyond, and
many residents I spoke with again exhibited a mixture of excitement at the
prospect of such a stern accountability mechanism and skepticism that it
will ever be exercised. Subsection H of this article also clarifies and extends
the conditions under which the other participatory mechanisms become
legally binding (*vinculante*).[25] For referenda and plebiscites, the threshold
is reached when at least one third of registered voters of the relevant geography participate in the initiative. For consultations, the threshold is only 15
percent of the relevant registered electorate. Interestingly, Article 32 leaves
the office and functions of the mayoralty intact and largely unaltered, even
retaining the forever-awkward and widely panned title, jefe de gobierno.[26]
Nevertheless, the constitution solidifies a great many other changes that
verge on the radical for Mexican society, especially some of the more progressive stances espoused in the Charter of Rights, the limits and checks
potentially imposed upon the executive branch and other elected powers, and the affirmation of the project of democratization and the further
empowerment of citizen participation.

Political reform and its ultimate products and effects represent the culmination of a conflict over the political geography of Mexico that developed
over the last century. The rapid growth of the city to unmanageable proportions in the latter half of the twentieth century, a series of political and

geological crises and a developing plurality within the PRI about how to solve them, and gross mismanagement and transparently corrupt adminis- trations conspired to create space for a leftist alternative to the ruling party and eventually a rupture in its control of the city and the country in the 1980s and 1990s. Incremental reforms passed from the 1980s onward and offered as concessions to the prized electorate of the Federal District even- tually proved insufficient to quell this energy, but it took both the presi- dential return of the PRI and the sanctimonious AMLO and his Morena compatriots to create the conditions for the Mancera/Peña Nieto alliance from which the 2016 political reform emerged. These reforms promised chilangos their long-awaited political emancipation from federal control, transforming the Federal District into the thirty-second state and creating a plan for the drafting of the city's first constitution. This document, which victoriously enrolls this emancipatory narrative with the fiery language of its opening passages, strikes out in progressive directions in its protections and guarantees, and even enshrines a collective notion of the right to the city. The alliance that created the new constitution and the manner in which it was produced, however, potentially portended the ominous return of the PRI's influence in the city even as the PRD struggled to maintain its own political and geographical base in the face of the increasingly potent chal- lenge of AMLO's Morena. Thus was the city made free of the country only by its leaders' recourse to the power of the federal state and the party whose rule had generated much of the energy behind this territorial conflict in the first place. Moreover, the grand assurances of the contract now offered de jure to the capital's residents were contested even before they acquired force, and the realization of their progressive promise will no doubt require countless difficult pilgrimages to the figurative Hyde Parks of legal and political authority.[27] Small wonder, then, that this momentous victory for the city's grassroots was tempered by a dubious, even melancholic air.

In the next section, I will explore and attempt to explain some of the complexity of the social context for this historical moment and the mean- ings attached to the political and territorial tensions that gave rise to it. The lens of revolution, I argue, is the single most important and appropri- ate filter for this analysis, for at least two reasons. First, and foremost, revo- lution—in various guises—was far and away the most common rhetorical and political frame put to work in pursuit of political reform, from elected officials, NGO leaders, and grassroots activists to cab drivers, bartenders, and amas de casa. It is, for instance, precisely the language and history that many of the most significant actors involved in political reform used to pro- mote their agendas, further their plans, and justify their actions. Second, it allows this historic moment and its social and political significance to be transected from a number of angles, producing an analysis that can at least begin to approach the rich mixture of often contradictory attitudes and motivations that have surrounded the process of political reform from its faintest murmurings in Mexico City.

VECTORS OF REVOLUTION: POLITICAL REFORM AND THE REVOLUTIONARY STRUCTURE OF FEELING

The concept of revolution enjoys a singular significance in Mexico City. It is an inestimably complex concept, burdened with histories that remain deeply disputed but nevertheless serve to elevate it to legendary proportions, and an almost mythic appeal as illusory in its promises, perhaps, as was victory in the Mexican Revolution that lends the concept so much of its peculiar flavor in the capital. As many have argued, revolution in Mexico should not be understood as a singular event, but rather as elongated, episodic, and plural, owing to the nature of its historical referents and the fractious character of the political forces claiming its mantel in contemporary struggles. Revolution is an idea that continues to animate politics across the country, to inspire strong feelings and drive cults of personality, and to legitimate actions and form the ideological basis of resilient movements that span the political spectrum and the national geography. Given this breadth of appeal and depth of expression, it is also an idea that allows its users to trace connections between a collection of events or practices that might otherwise be considered discrete, and to explain the coincidence of contradictory impulses that enrich one another even as they seek each other's negation. Political reform in 2016 and 2017 brought many such revolutionary impulses and implications to the surface of Mexico City's political landscape, and pushed contradictions that had simmered for decades into heightened intensity or extended geography, pulling in larger sets of national and international connections. Indeed, the process of political reform was, from the beginning, steeped in, shot through with, and in many ways carried along by the idea of revolution in various guises.

Neither is revolution a straightforward affair in the annals of critical, social, or political theory, Mexican or otherwise. From Arendt to Žižek, political theorists of the last century have spilled as much ink on this as perhaps any other topic, postulating and contesting all manner of spatial metaphors ([non-]linear, progressive, global, etc.) and temporal modes ([non-]reformist, permanent, continuous, etc.). In both Spanish and English, the concept evolved from rather auspicious origins among European meditations on the movement of celestial bodies. As Williams (2015, 209, original emphasis) has it, "In all its early uses it indicated a *revolving* movement in space or time." The notion of revolution as a break with history, "as bringing about a wholly new social order" followed evolving political usage as uprising (though more distinguished in its targets than the related rebellion) and fundamental change (Williams 2015, 212). For Arendt and others, the foundation of elliptical movement (in both common senses) retained a central structuring role. Revolutions are at once born of regenerative and reclamatory energies, bound to return to points of origin, and also generative of movement beyond mere orbital or axial rotation: "The revolutions started as restorations or renovations, and . . . the revolutionary pathos of an entirely new beginning was born only in the course of the event itself" (Arendt 2006, 27). But not every correction, insurrection, or revolt can be cast as revolutionary, Arendt insists. Adding an emancipatory normative qualification highly resonant in contemporary chilangolandia, Arendt (2006, 25) argues that "only where change occurs in the sense of a new beginning, where violence is used to constitute an altogether different form of government, to bring about the formation of a new body politic, where the liberation from oppression aims at least at the constitution of freedom can we speak of revolution."

Guided by and refracted through this multivalent understanding of revolution, this section will present a fourfold analysis of the rich palimpsest of feelings, motivations, and rhetorical postures that together form the conditions of emergence of political reform in Mexico City. Though metaphors abound for describing this mental and material terrain, it is best approximated by Williams's (1977, 132, original emphasis) notion of a structure of feeling, described as follows:

> The term is difficult, but "feeling" is chosen to emphasize a distinction from more formal concepts of "world-view" or "ideology." It is not only that we must go beyond formally held and systematic beliefs, though of course we have always to include them. It is that we are concerned with meanings and values as they are actively lived and felt, and the relations between these and formal or systematic beliefs are in practice variable (including historically variable), over a range from formal assent with private dissent to the more nuanced interaction between selected and interpreted beliefs and acted and justified experiences. . . . We are talking about characteristic elements of impulse, restraint, and tone; specifically affective elements of consciousness and relationships: not feeling against thought, but thought as felt and feelings as thought: practical consciousness of a present kind, in a living and

interrelating continuity. We are then defining these elements as a "structure": as a set, with specific internal relations, at once interlocking and in tension. Yet we are also defining a social experience which is still *in process*, often indeed not yet recognized as social but taken to be private, idiosyncratic, and even isolating, but which in analysis (though rarely otherwise) has its emergent, connecting, and dominant characteristics, indeed its specific hierarchies. These are often more recognizable at a later stage, when they have been (as often happens) formalized, classified, and in many cases built into institutions and formations.

The meanings and values that form the structure from which political reform emerged in Mexico City over the past several decades and more especially in the past several years, I argue, are most clearly organized around ideas of revolution, forming what I call a revolutionary structure of feeling. While Williams explains that such structures are more easily approachable posthumously, owing to the difficulty of understanding a structure in motion and in a constant state of becoming, this observation clearly varies with the temporal character of the structure in question, its rate(s) of change, and the geographical range of its inherence, among other factors. In the case of the revolutionary structure of feeling under examination here, it is worth noting that some of its deepest roots can be readily traced as far back as the earliest decades of Mexican Independence and the liberal reform(er)s of the 1850s, though I have confined my analysis to a window opening near the end of the Porfiriato. This temporal scope allows for a better approximation of structural contours than might be possible in the analysis of a structure of a comparatively narrower frame. This analysis is also pursued along vectors that cut through this revolutionary structure along paths contemporarily extant in the daily life and political theatre of the city, an analytical orientation that does not attempt to definitively chart the entirety of the structure, but rather to trace paths through it that effectively contextualize both the events and processes in question and the innumerable ways these are politically pursued, rhetorically refracted, and historically positioned and memorialized. Ethnographic observations, extremely helpful for making sense of a shifting structure of "meanings and values as they are actively lived and felt," are here blended with contemporary and historical media reports and secondary literature, in order to effectively trace through the structure along the four vectors I have identified. This analysis is not intended to offer explanations for the trajectory of political reform, per se, but rather to contextualize it in important and meaningful ways. Moreover, this analysis draws on and extends material and arguments presented in the foregoing chapters, which find a collective culmination of sorts in this revolutionary structure of feeling.

Examination of the first vector of this analysis will be propelled by the idea of the Mexican Revolution, the high point of which (the Constitution of 1917) motivated not only the specific timing of 2016–2017 reforms, including especially the constitutional process, but the orientation, rhetoric, and ideological underpinnings of many of its most significant players. The second vector will follow closely on the heels of the first, but will veer off to extend the notion of an urban revolution to the history explored above, wherein an increasingly politically distinct capital city disturbed by its radical disenfranchisement is emancipated from the federal state that dominated it for more than a century. The third vector will present another narrative of urban revolution, wherein Mexico City has been envisioned as the political and cultural vanguard of Mexico (often treated as a backward or underdeveloped country), popular in various iterations for more than a century among leaders from nearly every camp, from Díaz to AMLO. The fourth and final vector shares a similar understanding of Mexico's urban revolution, though again charting a distinct path through the structure. Rather than seeing Mexico City as a tool of or catalyst for modernization, globalization, or other vaunted bastions of international development, this fourth vector looks to its own, and proposes Mexico's urban revolution as a vehicle for the redemption of the promises of the many forestalled, failed, and forgotten revolutions of the country and the city.

VECTOR I: THE MEXICAN REVOLUTION AND THE CONQUEST(S) OF
THE CAPITAL

The first transect of the conditions of political reform is cut by the most obvious perspective of revolution in Mexico, by attempting to forge connections between political reform and the Mexican Revolution. This is a time-tested strategy for grabbing and maintaining legitimacy in Mexico, popular since the Revolution broke out in 1910. Always a complicated affair, the Revolution was, in the long view, less a conflict between an *ancien régime* and a liberating vanguard than it was a chaotic and episodic collection of rebellions and civil wars between regional factions of the armed forces, traditional caudillos seeking enrichment or social change, organized peasants and peons bent on agrarian reform, the sons of well-to-do families frustrated by their lack of prospects in the calcified Porfirian state, and the disaffected and growing middle sectors (see Knight 1986; Joseph and Buchenau 2013). Though successful in deposing the aging Porfirio Díaz, the revolt which promised to finally bring democracy to Mexico quickly devolved into a series of intrigues, coups, and civil wars that spurted flares until the PRI's consolidation of power under Cárdenas in 1934. This made telling a clean story about the Revolution a difficult task, but certainly not for lack of trying. In the century since its zenith, claiming to be the true inheritors or champions of the Revolution has become common political practice, as has the related practice of accusing opponents of Revolutionary betrayal. The PRI was only the most successful of a series of parties and movements that have ostentatiously employed and embodied such tactics, and its name correctly identifies its attempt not only to legitimate its political project of national rule and sectoral and regional integration under the flag of the Revolution but also to extend that Revolution's time horizon indefinitely. As the *Institutional* Revolutionary Party, its leaders claimed the mantel of the Revolution in perpetuity. As discussed above, however, PRI rule did not bring the democracy that the Revolution's first post-Porfirian President, Gustavo Madero, had promised. That task was later claimed by a breakaway PRD, whose name takes the next logical step, in claiming the Revolution as democratic.[28]

Despite the disparate claims of the Revolution's many combatants and camps, they shared an enemy in the autocracy of Porfirio Díaz. Under his control, the resplendent capital city came to dominate even the country's sleepiest regions through newly professionalized military and police forces and impressive infrastructural projects (especially railroads) that spanned Mexico's highly varied terrain (as discussed in Chapter 1). His reforms, bent on liberalizing and modernizing the country, created class and regional cleavages that would cement a general animosity between the capital city and the rest of the country, with grievances that emanated from dispossessed peasants in the central and southern agrarian regions to politically stunted professionals and *hacendados* of the arid and industrializing North (Joseph and Buchenau 2013). In a general sense, then, the Revolution brought together regional forces intent on curbing the power of the capital over their affairs. After Díaz's ouster, PRI control would reverse this trend as the city grew throughout the twentieth century. Though federal control remained seated in the capital city, the Federal District came to be politically dominated by a party whose strength was based in a federal electorate and fed by federal finances, even as the city's own exploding populace increasingly chafed at a party unbearably physically close but oppressively politically distant from its own concerns. The first conquest of the capital city—in the sense of belonging to or emanating from—was largely accomplished under Díaz, as the capital spread its tentacles across the uneven topographical and political surfaces of a poorly connected and highly regional Mexico. Subsequent conquest of the city, then, is conversely represented by the PRI's decades-long struggle to maintain and later reestablish control over its volcanic throne. This allows both sides of this antagonism to claim political reform as an end stage of the Mexican Revolution. For the national PRI, gifting more direct democracy to the capital city fulfilled the greatest of *Maderista* promises on the one hand, and on the other took away the special privilege that had belonged to the Federal District by giving it equal federal status among the other thirty-one states of the Union (setting aside the question of the party's war of position with respect to the process). For the PRD, self-determination and emancipation from federal oppression represented both the democratic fulfillment of the Revolution and its final act of liberation. It is for precisely these reasons that Peña Nieto's reforms specified January of 2017 as the delivery date for the city's constitution, so that its completion would fall on the centennial of the national Constitution of 1917 that nominally ended the country's last great conflict; a highly convenient political bookend to a century of Mexican Revolution. Though the initial public rollout of the 2016 reforms was decidedly awkward (as discussed above), by the time my initial fieldwork ended in the early summer of 2016, the official pomp anticipating the January constitutional arrival amidst the centennial was already swelling to sticky proportions, and, given its (however ambiguous) support among the major parties, even the dubious Morena, spilled into nearly every corner of the city. In the months and years that followed, I have routinely noticed or been cynically nodded toward faint and shabbied but unmistakable visual evidence of this official zeal lingering on the city's vertical surfaces. This too, is in some ways reminiscent of Celorio's (2004, 49) description of the official aesthetics of nearly a century before, when, "Once in place, the Revolution plastered the entire history of Mexico over the walls of public buildings."

VECTOR II: THE URBAN REVOLUTION AND THE EMANCIPATION OF THE CAPITAL

In the second vector, revolution hews closely to the popular PRD and later Morena understanding of political reform as the emancipation of the capital from federal domination, in all the ways discussed above. At its point of incision, however, this vector distinguishes an *urban* political revolution from the national revolution explored through the first vector. Here, the lines of conflict are temporally located not in the tensions of city and country first established in the grand modernizing projects of the Porfirian

Capital, but rather in the explosive decades of the second half of the twentieth century. From this perspective, political reform is seen as the culmination of popular efforts by urban citizens themselves, and the heroic figures have tended not to be the heads of political parties, the operators of great machines, or the generals of victorious armies, but rather the organic intellectuals and everyday residents of the city's marginalized zones and populations, and to a degree the collective political subject the capital city has become (though with at least one important caveat, discussed below). The urban revolution, in this line of thought, is not simply a stage of an established political revolution, but rather is a revolution unto itself. From this perspective, this urban revolution consists of a geographical and temporal evolution of Mexico City beginning around midcentury, which only partly overlaps with the developments of political geography and the machinations of PRI and PRD regimes discussed above. My analytical use of the phrase *urban revolution* here coincides with the Lefebvrian theory of that name in one key respect especially: the notion of a quantity/quality change, a critical point in Mexico City's development. As discussed in Chapter 1, the city's peripheral growth from the 1940s onward created new governance challenges to which successive PRI administrations failed to adequately respond. The resultant growth of popular resistance in marginal social and political sectors and peripheral zones eventually changed the political tenor of the city as an aggregate and fostered growing calls for urban autonomy. The political realization of this quantity/quality change, however, was not simply a process of crude aggregation. Sharing Dikeç's practical focus on (2013, 78) "the doings rather than the beings of political actors," the urban revolution in the second vector emphasizes how the experience and effects of explosive growth were and continue to be understood and experienced as a *political* transformation by and for the capital's residents. This is perhaps especially evident among the city's newest—most revolutionary, in this sense—inhabitants, its twentieth-century rural-to-urban migrants, *paracaidistas*, and their progeny, about whom certain developmentalist ideas continue to deeply influence mainstream contemporary attitudes toward the urban poor.[29] As discussed in Chapter 4, theories of ruptural and inaugurative (i.e., revolutionary) politics often narrow their temporal opportunity to moments that are stubbornly indeterminate in theory but nearly invariably razor-thin and fleeting in practice. As with the defeat of the Corredor project, such moments are nearly impossible to predict, and even difficult to imagine before they are proven in practice. The notion of an urban revolution that animates contemporary politics in this way, however, insists on a rupture slower

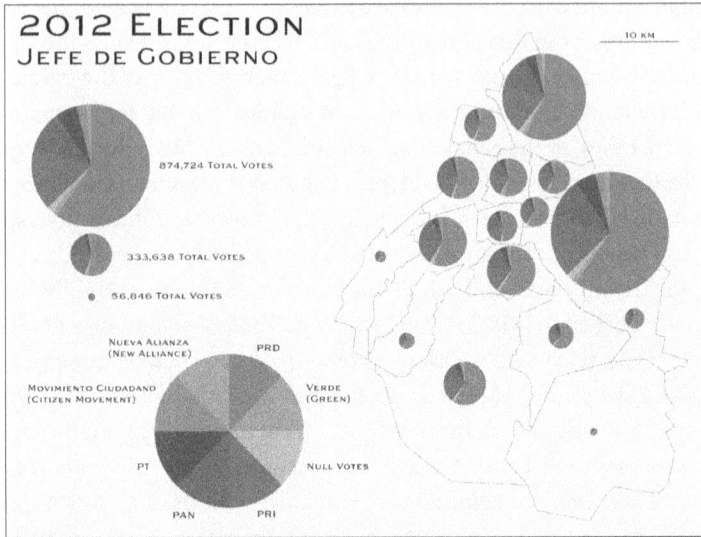

FIGURE 5.11. Distribution of votes by party, mayoral election (jefe de gobierno) of the Federal District. The electoral pie charts are scaled to the number of total votes for each delegacion, and show a strong showing for the PRD (darker blue), followed by the PRI (red) and PAN (purple). Data sourced from IEDF, cartography by author.

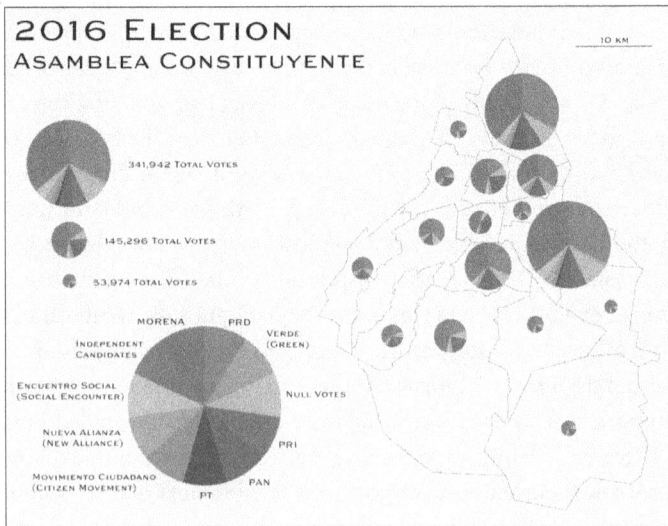

FIGURE 5.12. Distribution of votes by party, Constitutional Assembly election of Mexico City. The electoral pie charts are scaled to the number of total votes for each delegación. Unlike the election results depicted in Figure 5.11, these data illustrate the rise of Morena (purple, upper left) at the expense of the PRD (darker blue, upper right). Also notable is the large number of null or empty votes in this election, which represent a kind of active abstention. Data sourced from IFE, cartography by author.

in coming, enabled by a quantitative massing that rose like an unforeseen tide born of a political-economic gravity both seated in the intimacy of the capital yet imperceptibly permeating and steadily agglomerating the country's farthest reaches of national patrimony. Though chilangos may speak of the city in this way (as a kind of collective agent arising from this demographic transition), room seems always to be elbowed in for significant difference, along the lines Williams (2002) was expounding contemporaneously to this change.[30] In this way, too—exploding the temporal moment beyond an inaugurating instant—this urban revolution eschews a sharp zenith of parabolic growth, a specific numerical tipping point or threshold, in its transformations. Rather this revolutionary metamorphosis, as Dikeç (2013, 81) explains of the Arendtian concept of plurality, "is not a mere numerical matter."

Revolution in this second vector is therefore inescapably a matter of urban geography. As the Mexico City Charter for the Right to the City makes clear, however, the geography in question after the Miracle decades could no longer be the Federal District, but a metropolitan region with issues increasingly beyond the reach of the GDF. In this respect at least, political reform appears at best inadequate, as the political calculi of states and parties in 2016 fumbled or conspired to birth the spirit of a new Mexico City into the frail body of a Federal District already cracking under the weight of its metropolitan morbidity. A second geographical consideration in this vector revolves around the rise of AMLO and Morena, even before 2018. Morena has grown rapidly since its emergence in 2012, garnering an increasing share of the vote in several subsequent elections (see Figures 5.11 and 5.12). A good deal of Morena's urban base is made up of former PRDistas, many of whom feel betrayed by Mancera and by the national party organization more generally.[31] For many, Morena's ascendency itself represents an urban revolution far more than any that could result from the duplicitous dealings and shifting alliances of the PRI and PRD solidified in the Pacto por México.

AMLO's PRD bids for the presidency in 2006 and 2012, both of which ended in defeat, represent the urban revolution as it might have been, and his 2018 victory under the Morena flag its ultimate fulfillment. For his part, the triumphal rhetoric of his Cuarta Transformación (Fourth Transformation) places his government at the culmination of a linear revolutionary heritage populated by Mexican Independence (1810–1821), Reform (1858–1861), and Revolution (1910–1917). Though many have criticized AMLO's populist tactics and the cult of personality that seems to flavor much of his following, and though the country and capital remain bitterly divided over his leadership (as of 2022), Morena's supporters nevertheless see in their party and its leader a qualitatively different orientation than that of the PRD and other leftist opposition parties. As Figure 5.12 illustrates, Morena's support has been strongest at the city's peripheries, and its name and much of its rhetoric position it as the party of traditionally marginalized groups.[32] This gives Morena considerable clout as the party of the urban revolution, as its political geography and rhetorical positioning align it with the areas and populations that produced the quantity/quality change identified as the urban revolution from this perspective. Insofar as AMLO and Morena have captured this geography and this revolutionary (transformation) narrative, the emancipation of the capital from federal control and emergence as political subject have now become deeply entangled with the national ascent of the city's former mayor and his party's takeover of the city's political geography.

VECTOR III: THE URBAN REVOLUTION AND THE VANGUARD CITY

A third vector likewise follows the idea of a distinctly urban revolution, but from a different vantage. From this perspective, Mexico City acts as something like a vanguard agent in a process of revolutionizing a Mexico seen to be lagging behind. This line of thinking is perhaps the oldest of the four perspectives presented here, and its roots can be found farther back in Mexican history than even the liberal reforms of President Benito Juárez García and his contemporaries. Indeed, attempting to modernize Mexico by way of the capital city is a strategy that has historically cut across a number of practical and ideological divides. For Emperor Maximiliano, the erection of grand boulevards like the Paseo de la Reforma (renamed after Maximiliano's execution) and other beautification projects were pursued in an effort to reposition Mexico aesthetically with the capitals of Europe, bringing urbane culture to a backward agrarian country and reviving (or surpassing) the glory of the old colonial viceroyalty. For Porfirio Díaz, industrialization and modernization began and were pursued most vigorously in the capital, whose lingering aesthetic vestiges of this transformative era now form the substrata of equally vigorous processes of neighborhood change (see Chapter 4). For the PRI and its Revolutionary predecessors, the city was the hearth from which the various waves of the "state-sponsored cultural revolution" emanated (Joseph and Buchenau 2013, 109). These included a national education program oriented toward dampening the startling levels of illiteracy in the countryside and the commissioning of murals and other public art works that would educate and inculcate the populace with the spirit of the Mexican Revolution. These and other programs begun in the early years of the PRI were also always about forging a distinctly Mexican nationalism and civic culture, and, whenever possible, to tie the ruling credentials of the party and its leaders to the Revolution (as in the first vector). Mexican muralists of the early post-revolutionary decades would come to enjoy a global appreciation as an artistic vanguard (and a complicated relationship to both the Revolution and the state), "help[ing] to transform Mexico into an avant-garde centre of art" (López Orozco 2014, 268). As argued above, both the PRD and Morena have continued in this posture toward the capital, cultivating political bases in the city where their (center-)leftist rhetoric and practices find the greatest degree of sympathy and support. Social media and cultural commentary debates—particularly heated on Twitter, Facebook, and the comments sections of VICE and Medium, and related publications in the spring of 2017—both lauded and lamented the degree to which Mexico City's art, food, fashion, and other aesthetic scenes increasingly draw international attention to a country the rest of the world so often seems to regard with a disinterested mixture of fear and forgetfulness.[33] This aesthetic vanguardism is not the exclusive pursuit of connoisseurs of the city's fanciest fashionista parties, international music festivals, and hip art galas like the biannual Zona Maco, however. It has also been the explicit strategy of various iterations of the GDF for many decades, taking forms from Mancera's pursuit of the Corredor project to Regente Carlos Hank Gonzalez's attempt to turn Santa Fe's garbage dump, abandoned sand mines, and informal housing into his own Manhattan at the edge of the Federal District (see del Carmen Moreno Carranco 2008). This is also the strategy long at the heart of the push for the new international airport (scrapped again by AMLO in 2018), which boosters claimed would make the city the new regional hub for the Americas and the Pacific Rim, and allow Mexican industry and capital to forge stronger ties with Asian and American partners, in addition to making the country as a whole more globally competitive.

Constituyente and frequent constitutional apologist Porfirio Muñoz Ledo often explained political reform in terms that foregrounded the city's role as the cutting edge of progressivism and democratization in Mexico. In early 2016, he argued in an interview with *El País* that the Constituyente needed to negotiate a path between the PRI's attempts to limit the capital's autonomy and the PAN's ideological objections to the Assembly's progressive stance on social norms, including especially sexual and reproductive rights (Beauregard 2016). One of few political figures able to forge a productive working relationship with both the PRD and Morena, Muñoz Ledo's

faith in a leftist urban coalition to work through the Constituyente exemplified this line of thinking with regard to the city's capacity to act politically as an urban vanguard. The leftward drift of the city—even in the face
of a fractured political left vying for supremacy and control—enables it to
perform this role for Mexico. This is also a common perspective taken by
international observers, for whom the progressive nature of the constitution
demonstrates the potentially transformative role of the new entidad federativa within the Mexican Union.[34] Indeed, the temporal and geographical
coincidence of reform at every level in the 1990s in particular (even allow-
ing for disparate conditions) can also be seen productively in this light, as
Ziccardi (1996) demonstrates.

In this third vector, then, the urban revolution represents an opportunity
to use the moment of political reform to assert a new, urban-led program of
broader political change. Part of the work, as Marco Rascón insisted above,
lies in the mental labor of imagining alternative futures. Advocates of the
Mexico City Charter for the Right to the City likewise note the importance
of seeding ideas into the urban consciousness. Recognizing, however cynically, the necessity of this imaginative work, Knight (1986, 527) concludes
their massive two-volume tome on the Mexican Revolution with this question: "It must be admitted that the real developments of the Revolution did
not in any way resemble the enchanting pictures which created the enthusiasm among its first adepts; but without those pictures would the Revolution
have been victorious?" Such "así sí" convictions belong to the theoretical
side of the material whole that is Marxist praxis, a central preoccupation of
twentieth-century Mexican philosophy.[35] Carving out fresh spaces of political imagination in this fashion positions such everyday activity as a kind of
vanguardist revolutionary praxis. But as McAdams (2017, 60) has recently
argued, even the theorist par excellence of the revolutionary vanguard,
Vladimir Lenin, "thought of it in primarily revolutionary terms" rather
than as later adapted to the complicated business of state. Lenin's vanguard,
on this view, was more for the fight than the future, an instrument of revolution rather than an infrastructure for the aftermath. This qualification finds
a dour corollary in Mexico's dubious history of failure to deliver on revolutionary promises, a lived reality never far from the minds of the capital's
residents. After all, the 1917 constitution was itself once heralded a triumph
of social progress. As Cuauhtémoc Cárdenas's rather sober keynote address
above so crucially recalled, the space between de jure and de facto is best
approached as dangerously unsounded terrain. In the pregnant moment of
political reform, it's also worth noting that that space also harbors other,
competing interests. Alongside and behind expanded reproductive rights,

greater protections for indigenous peoples, and novel mechanisms for political transparency lurk the forces of global finance and throngs of expatriates and their eager spirit guides from among the internationally minded, the upwardly mobile, and the (would-be) nouveaux riches. Rights talk, so central to the victories of political reform understood in this way as urban revolution, is also perfectly tailored for smuggling in, under the banner of progress, the regulation (and, especially in the case of Mexico City, regularization) so coveted by the forces of that other meaning of Lefebvrian urban revolution, wherein industrialization cedes to urbanization the dominant role in a thoroughly globalized process of accumulation.[36]

VECTOR IV: THE REDEMPTIVE REVOLUTION

In a fourth vector, revolution is endowed with a special capacity for redemption. Specifically, I will here focus on understandings of the moment of political reform as endowed with the potential to redeem the promises of revolutions past. In the common parlance of the residents of Mexico City, the utterances and attitudes that belong to this perspective are often expressions of hope, or even better of faith. To be sure, there is no small measure of religious influence on this conception, despite the secular professions of many of its contemporary urban chorus; nor can the expansive and complex relationships with death now so (in)famous on both sides of the Río Grande be discounted in this formulation.[37] Like Benjamin's messiah ([1968] 2007, 255), whose vocation must be pursued in historical reiteration, the urban revolution too "comes not only as redeemer, [but also] as the subduer of Antichrist." "Only that historian," Benjamin writes, "will have the gift of fanning the spark of hope in the past who is firmly convinced that *even the dead* will not be safe from the enemy if he wins" (255, original emphasis). But the role of revolution in this vector is distinct from that of a conventional messianic figure, as the urban revolution (for which political reform here stands as proxy) is responsible neither for transforming something otherwise intrinsically evil, per se, nor for forging a divine connection, but rather for reviving and fulfilling the promise of previous revolutionary moments, events, and impulses, regardless of their original outcomes. As in Arendt's and Williams's formulations, revolution in this vector signifies both the emergence of a radically distinct present and the simultaneous reinscription of the past. For those chilangos for whom this understanding has meaning—whether or not they phrase it precisely in this way—the moments to which revolution as a redemptive exercise or ideal return are also multiple, including revolutionary histories as experienced, as recorded, as understood, and as otherwise imagined.

Such forays into the past are as commonplace in Mexican thought as they can be in daily conversation. In a marvelous introduction to Mexican philosophy of the last century, for instance, Sánchez and Sánchez (2017, xxx, original emphasis) argue that "Mexican philosophy is not an accomplished fact," but rather should be defined in part by a driving reflective impulse, which "persistently returns to (i.e., *reflects back into*) the Mexican circumstance as its ground."[38] They stress that this reflexive impulse is critical in nature, and resists the purely nostalgic. Carlos Monsiváis (1997, 100), one of the capital's most intrepid and keenest observers and most incisive cultural critics, saw a similar multivocality at work in art of one of Mexico's most famous performers, the comedic actor Cantinflas:

> Cantinflas is a recapitulation and a point of departure. The trousers held up beneath the waist were already a commonplace of the comic strip (*Chupamirto*) and in popular neighborhoods, speech encoded into the absurd (piss-take as a vocation and the confession of ignorance) abounded. But Cantinflas's style perfects and endows nonsense (the failure of eloquence) with humor and, in the entanglement of endless broken sentences, provides a glimpse of a frankly urban mentality, without rural comparison or allegory.

As understood by Monsiváis, Catinflas not only embodied the marginalized experiences of what I have labeled Mexico City's urban revolution, but performatively created space for their critical reinterpretation and creative reinscription into different logics, for different ends. Within this fourth vector, however, previous revolutions are not simply reanimated, jocularly or otherwise. The urban revolution, in this line of thinking, is carried by sober conviction toward the redemption of the departed. In Sophie Fiennes's (2012) film *The Pervert's Guide to Ideology*, Žižek offers a resonant vision of such spectral revolution:

> In revolutionary upheavals, some energy, or rather some utopian dreams, take place, they explode. And even if the actual result of a social upheaval is just a commercialized everyday life, this excess of energy, what gets lost in the result, persists. Not in reality, but as a dream, haunting us, waiting to be redeemed. In this sense, whenever we are engaged in radical emancipatory politics, we should never forget, as Walter Benjamin put it almost a century ago, that every revolution is not only—if it is an authentic revolution—is not only directed towards the future, but it redeems also the past, failed revolutions. All the ghosts, as it were, the living dead of the past revolutions,

which are roaming around unsatisfied, will finally find their home in the new freedom.

This perspective would have it that as the true revolution, the urban revolution in Mexico City reveals and reanimates a parade of previous revolutionaries and their visions of the promised land not as they were, but as they could and should have been. The moment of political reform, for many, held just such lofty possibility.

There is, perhaps obviously, a great deal of historical overlap between this vector and each of the others, in the sense that this take on revolution includes events, processes, (etc.) that figure centrally in those vectors. Where this redemptive take on revolution differs, however, is the role that revolution plays in this tracing through the structure of feeling that surrounds political reform, and thus how such elements are understood through this lens. For example, while the urban revolution considered through the second vector may have it that the city's emancipation is the culmination of a democratic revolution begun in 1988, from the perspective of this fourth vector it is the urban revolution that saves and redeems the promise of this democratic revolution, which may or may not have failed. It would be easy to call the PRD's democratic revolution a failure, as many have and do in the city, given its well-documented history of fraud, corruption, and betrayal, to say nothing of the defection of its highest-ranking officials and its most politically and symbolically important bases of support in the city. The lingering hope that can be found among supporters of political reform, while acknowledging these failures, nevertheless affirms their value and finds in the city's new political autonomy the opportunity to realize the promises of the democratic revolution in whose name they once fervently strove. The urban revolution, on this view, takes the democratic revolution to a higher plane, reviving its energy despite its practical faults. The same can be said about the politics of the revolutionary youths whose tragic deaths at the hands of the PRI government in 1968 remain an open wound for many of the city's residents, the pain of which is revived and reinvigorated with each new discovery of a mass grave or story of missing students or murdered journalists.[39] While the student massacres did not fully galvanize opposition to the PRI nor bring about the social and cultural revolution many undoubtedly hoped (and died) for, placing their struggle in the lineage the 2017 constitution is in part an attempt to redeem their sacrifices and claim victory for many of their dreams.

Speaking in this fourth revolutionary register undoubtedly obscures as much as it illuminates. Working back through my fieldnotes also serves as a

reminder that its voices can often be difficult to hear, decipher, and understand. Many of the expressed sentiments and affective evocations pertaining
to this vector represent what I have elsewhere called "the hopeful refusal of
perceived political realities" (Gerlofs 2019, 383), a wordy description aimed
at understanding an irreducibly complex and multiple behavior or convey-
ance, which easily finds a place both in a long history of thought in and
about Mexico and in the quotidian atmosphere of the capital. Revolutionary sentiments of this variety routinely register as undercurrents of one sort
or another. This could certainly be said to be the case with respect to the
constitution's recognition and protection of Indigenous communities in the
city. Though much of the cultural revolution propagated by the PRI and the
revolutionary state more broadly intended to promote a vision of Mexico
united by a mestizaje identity that celebrated the mixture of indigenous
and European heritage, the work of some of the era's most celebrated artists
sharply contests this harmonic narrative of post-Revolution Mexico, as do
the many aspects of everyday life in the city still visibly and palpably saturated by racial hierarchies.[40] The new constitutional language, which seeks
to affirm and defend these populations without subsuming them under a
nationalist integration project, arguably demonstrates a redemption of the
Revolution's ostensible aims.

There are analytical risks in following such a vector, however carefully,
that should be attended. Crane (2015, 5), for instance, sees a danger in reifying the static and uncritically mythologized opponent of a PRI state unintentionally and unreflectively constructed through "kaleidoscopic narra-
tions of genetic continuity in a movement family" (the lumping together
of distinct movements across diverse conditions and against different iterations of the PRI state), which he argues "reflect what Ross . . . names 'a
police conception of history.'" Certainly, there is merit in this concern for
the distinct conditions and characters of diverse movements and the governments they oppose, not least on the tactical plane as these events unfold
(and on the analytical plane on which such overcoding obscures them).
But as Joseph and Buchenau (2013)—among a plethora of others—have
amply demonstrated, contesting and claiming this history is an incredibly
significant component of contemporary Mexican politics, a fact not lost
on many of its diverse combatants.[41] This role of aggregating decades of
activism under a new heading is one of the ways many now understand
and articulate the relevance of political reform, as a redemptive moment
in which the concerns of women, minorities, workers, peasants, and others whose own unique movements have had greater or lesser moments of

triumph and defeat over the course of the last century may find a rebirth of sorts under the banner of political reform as an urban revolution. For the AMLO faithful who kindly shared their aspirations with me at the women's march in 2016, this promise seemed a fundamental source of enthusiasm. Even holding aside the questions of the Cuarta Transformación that was to commence some two years later, political reform and the city's constitution represented, as had Monsiváis's Cantinflas, both "a recapitulation and a point of departure."

These four vectors illustrate some of the contours of the most salient frame through which chilangos and outsiders pursue and understand the revolutionary changes bound up in the processes of political reform. While they do not themselves provide many opportunities to attribute causal or even catalytic powers, they demonstrate the revolutionary milieu from which particular packagings of political reform have emerged in recent years, and from which they continue to be drawn in defense of the constitution or as entreaties for greater civic engagement in its wake. This revolutionary structure of feeling therefore provides residents and observers the raw material from which may emerge their imaginings of Mexico City's political reform as an end stage of their long national Revolution, as an urban revolution unto itself, as the vanguard of an urban-led revolution for Mexico, and as the redeemer that can call forth and reanimate the skeletons of all the fallen heroes and make good on all the promises of the many revolutions that preceded and foretold its glorious arrival. This milieu also makes it possible for contradictory positions to coexist within parties, households, and persons. It is for this reason that a chilango can casually retort—with a look of surprise that betrays a slight suspicion toward my lack of grasp on the obvious—that while it is of course an elaborate trick of the cruelest kind, political reform nevertheless represents a moment for hope in the great potential of the process.

CONCLUSION: THE WEIGHT OF REVOLUTION[42]

The year 2017 was special in Mexico City. It was a year for remembering the great deeds and legendary heroes of the Mexican Revolution, and for reasserting and reestablishing the principles for which it was allegedly fought. I returned to Mexico City in April, and within the first few days of the month I was able to attend two exceptional exhibits at the Museo de Arte Moderno and the Palacio de Bellas Artes, both of which were organized for this centennial year and dealt with the Revolution's aftermath. At the Museo de Arte Moderno, *Escenarios de identidad méxicana* (Scenes of Mexican identity) blended portraits and landscapes of peasant and village life with scenes of volcanic eruptions and violent encounters with mythical beasts in tracing the process of the formation of national identity from the end of the Revolution through the 1980s. At Bellas Artes, *Pinta la Revolución* (roughly, Paint the revolution) presented a no-holds-barred visual examination of the many depictions of what were, in fact, many revolutions, from stark and moving paintings of bloody encounters to cynical satirizations of parties and Revolutionary figures. Both exhibits featured work by the most important figures of twentieth century Mexican art, including Diego Rivera, David Alfaro Siqueiros, and José Chávez Morado. Moving through both exhibits, I was struck once again, as I had been a year before amidst the energy of the constituyente, by the complex and contradictory feelings that surround the idea of revolution in Mexico City. The weight of the Mexican Revolution, with all of its progressive and democratic ideals, its wonton destruction of rural villages and grand urban architecture, its nearly two decades of political intrigues and assassinations, and the lingering wounds which still cut deeply through the landscape and national psyche, exerts pressure on the contemporary processes of political reform in ways both seen and unseen.

In the wake of Peña Nieto's 2016 reforms, several decades of incremental political reform finally culminated in the firm establishment of the city's autonomy and its recognition among the states of Mexico. The legal process of political reform, and even the heated debates regarding the contents of its first constitution, were fairly straightforward. The city's mood surrounding these developments, however, was anything but. The very same people who thanked heaven for the arrival of the long-awaited reforms were often quick to caution that their passage was only part of an elaborate and sinister political scheme, while those who viciously derided Mancera and the PRD for their unholy alliance with the PRI and Peña Nieto would nevertheless insist on the hope they found in the most important prize of the Pacto por México. The set of feelings, motivations, thoughts, rhetorical postures, meanings, and values most helpful for providing adequate context for political reform, I have argued here, center on ideas of revolution, and are collectively presented here as a revolutionary structure of feeling. Four perspectives on revolution, from the Mexican Revolution and its political uses to the notion of a revolution that redeems the promises of its predecessors, cut vectors through this structure in ways that illuminate the weight of history on the present moment.

This analysis demonstrates the complexity surrounding political reform as a revolutionary moment, and illustrates why there was (and remains) every reason for chilangos to be both apprehensive and excited about this process. Some of the most aggressive social protections and radical notions of collective ownership of the city have now become law, and the city is finally free of the clutches of a federal state whose political domination and sometimes violent subjugation spanned a century at least. But Mexico City is a better place than most for learning the hard lesson that the law is, above all, a relationship. The last century bears this witness with countless fraudulent elections and otherwise tainted exercises of cruelly anemic democracy, and the unfulfilled or unevenly attended promises of a Revolutionary constitution also once hailed as impressively progressive. Moreover, the city's constitution will require a bevy of supporting legislation and jurisprudence for its aims to be realized, and the 2018 elections were deeply unkind to the parties of the Pacto por México. Even the political emancipation of the city so greatly prized by chilangos left the old Federal District borders intact, ensuring at least a temporary stasis in regional political tectonics and keeping the powers of the city in check, but also holding fast the glaring inadequacies of a mismatched political geography a half-century at least in the making. To navigate this political moment, residents of Mexico City look to the stormy decades of their national Revolution a century before, which bequeathed them a steadily simmering cauldron of conflict embodied in a monstrous metropolis nestled into the volcanic Valley of Mexico and fittingly nicknamed the Lake of Fire. That violent and heady epoch of Villa, Zapata, and Díaz, and a hundred years of contesting its memory, left a great many lessons for contemporary chilangolandia. Sometimes even the most sincere of promises go unfulfilled, at least for a time. A moment of revolution can easily sprawl to encompass a century. And death of all sorts, to borrow the phrase, can be greatly exaggerated.

Conclusion

Any theory of the city must be, at its starting point, a theory of social conflict.
Manuel Castells (1983, 318)

When I concluded the fieldwork that informs the bulk of this book in the summer of 2017, developments in Mexico City warranted a certain amount of hope for the future of both the city's governance and its grassroots. Of course, that future was, and remains some years later, profoundly uncertain. As many expected, AMLO's Morena made a forceful showing in the 2018 elections both nationally and locally, bringing the former jefe to Los Pinos (though, true to form, he never resided there, opting instead to open it to the public as a museum) and Claudia Sheinbaum Pardo to the Jefetura. The race had finally been, at long last, his to lose. The rechristened capital has taken its place among the states of the Mexican Union, and has seen the first constitution in its history enter into force. Even before the onset of 2020's global COVID-19 pandemic, which has ravaged the city and the country as it has so many others, however, the honeymoon for AMLO and Morena was already in, or even past, its waning stages. Both Morena executives now face significant popular opposition even beyond traditional ideological chafing, owing not only to the their handling of the pandemic, but also the country and capital's political economy (from unpopular managerial decisions such as the 2018 cap on public sector salaries to broader problems like those evidenced by the second scrapping of the new international airport, which many saw as misguided at best and anyway oddly undemocratic, and the tragic, deadly collapse of a section of the city's perennially controversial and chronically unsound Línea 12 Metro line) and security (from AMLO's controversial formation of a military Guardia Nacional (National Guard) to combat the alarming rise in violence throughout the country during his first years in office to the embarrassing October 2020 US arrest in Los Angeles

of General Salvador Cienfuegos Zepada, the country's former defense minister, on charges related to drug trafficking), among numerous other concerns.[1] Add to this that even after the global pandemic subsides, whatever
new normalcy may await the city's residents has yet to be determined. In the
preceding five chapters, I have endeavored to contextualize this moment,
and to explore and explain some of the complex history and geography that
underlie, structure, and otherwise inform its appearance and reception. In
this conclusion, I will briefly present three additional lessons on the path
and pace of urban change that have emerged through these efforts, demonstrating how the insights gained from this study of Mexico City's urban
revolution both inform and challenge contemporary thinking and practice
on and in the urban age.

The first of these lessons on urban change insists upon a simple but
radical revision of the standards used to assess the world of urban politics.
To extend a Latourian concept, Mexico City's century of urban revolution
illuminates the blind spots of many commonplace trials of strength used
to adjudicate between success and failure for urban movements, to detect,
qualify, and activate latent political energies, and even to adequately parse,
in a political sense, life and death.[2] A somewhat rigid posture of this sort is
evident in social movement theory at its genesis in Castells's seminal tomes
The Urban Question and *The City and the Grassroots*, in which the moniker
of "urban social movement" is granted only to those "system[s] of practices"
that achieve the loftiest of revolutionary aims, or that, at the very least, articulate a demonstrable trajectory pointing "objectively towards the structural
transformation of the urban system or towards a substantial modification
of the power relations in the class struggle" (Castells 1977, 263).[3] If a movement achieved structural change, its recognition is secure. If not, well, perhaps not. Understandably, similar rigidity can also be seen in trenchant
disinterest among urban theorists and scholars in the notion of the right
to the city which they considered politically (and therefore conceptually)
compromised, or among activists for whom the idea felt passé or politicians
for whom it carried too much unwanted political baggage. The idea, and its
politics, were given up for dead. But as the foregoing chapters betray, even
death is not, strictly speaking, a permanent state of affairs. The political
leviathan was invited back to the capital by a beleaguered national populace dispossessed of their fears of its torments by exposure to the violences
unleashed by the heralds of the democratic transition and inheritors of the
PRI's gnarled state. Crestfallen figures like former jefe Marcelo Ebrard and
suspect PRI apparatchik Manuel Bartlett have returned to national prominence (Ebrard as the powerful Foreign Minister under AMLO). The new
international airport was begun, scrapped, renewed, nearly completed, and

scrapped again. And the right to the city was celebrated, relegated to the back bench of urban politics, revived and vigorously promoted, and ten years later finds itself enshrined in and empowered by the new legal foundation of the capital. Winning this right to Mexico's urban leviathan required a keen sense of the potential of this political geography and the fortitude to look beyond loss, betrayal, and a dormancy many confused for demise.

Mexico City's new right to the city is also a product of strategic political alliances made at crucial historical junctures. The highs and lows of this history suggest, as a fairly obvious observation, that grassroots movements that partner with the state seem to find dangers and benefits in near equal measure. While some members of civil society were always wary of involving even what seemed the most trustworthy of government partners, those that eventually became part of the CPCCMDC chose the recognition and power that came with the partnership of Ebrard and the CDHDF, among others. While this put the right to the city on the map in a way that it simply had not been before, it also created baggage for the concept and for the coalition, linking both to a particular regime and a particular political moment. When this moment faded and a new administration was installed, the concept was seen as outdated and non-pertinent to the new mayor's priorities (not to mention that its proponents now belonged to a political constituency that had no working relationship with the new mayor nor with much of his staff). This history of mixed results weighs heavily on contemporary conversations in civil society, and is one major factor separating various organizations and popular factions. That the right to the city explicitly found its way into the city's new constitution, and even more that the principles elaborated in the Mexico City Charter for the Right to the City so clearly influenced large sections of the constitution's Charter of Rights, however, suggests that even political moments that are short lived, even those perhaps perceived as failures can have lasting effects, including becoming the scaffolding of future political movements.

This argument is reminiscent of those of Mitchell (2003) and Holston (2008), for whom the legal establishment of a right can create a platform upon which a politics can be constructed. This fixing of a field within which a politics might occur is thus a foundational effect and objective of rights talk. What my analysis suggests, however, is that even a right which is recognized as a guiding principle but not given legal authority can perform such a role, in due course. For example, it took Enrique Ortiz Flores, Jaime Rello, and their many partners more than twenty years to earn any formal recognition for the right to the city. And when it was finally granted, it came only in the form of public endorsement, not legal authority. The Charter had no legal standing, and its political utility faded quickly in the wake of its public unveiling. Still, the document's publicity secured it a place in the political realm it had not otherwise enjoyed, and, though its position was legally weak, and though its most valuable partners perhaps wished to use it to accomplish distinct ends, its prominence enabled its later elevation to the status of law. Even this placement guarantees nothing, but does provide a sure footing for the construction of a potential politics based on the right to the city as a collective and complex right in Mexico City. The gradual and incremental process of political reform similarly illustrates the potential of such an approach, whether strategic or not. Though the earliest conversations regarding political reform sought to obtain complete autonomy for the Federal District, it took some thirty years for this to be legally realized. Along the way, however, defeños were given a representative council (though at first its role was purely advisory), an elected executive, a legislative authority, and elected local officials, all via incremental reforms. This process was halting, and required circumstances and interests to align in specific ways to create advantageous conjunctures in which the demands of the capital's residents might be fruitfully pursued.

The lesson in all of this is that the pace of urban change can appear chaotic, multiple, and difficult to precisely qualify. Reading this historical geography has proven this lesson many times over, and exploring this terrain ethnographically has taught me to tread carefully on the resting places of movements, machines, and ideas. In concluding his *The New Urban Question*, Merrifield (2014, 131) states this lesson in simpler terms, beginning with the often incredible paths of change:

> That's the hopeful side of stuff. The negative side is that all revolts, all social transformations you believe may have won one day can in fact transpire as Pyrrhic victories the day after, or the day after that. You realize that actually the social movement has transformed into something else, that it isn't exactly what you'd expected, even though it's a punctuating encounter in itself; something's happened which isn't what you intended. And there's no way anybody can know this to begin with, prior to engaging in struggle, no secret or double agent, no great escaper or great refuser. You just have to deal with it and see what happens, use foresight and insight, experience and ingenuity, activity and acumen, even as you acknowledge that there's no recipe book, no simple magic formula.

The second lesson relates to strategic concerns, and is best approached through the political and theoretical concept known as the non-reformist reform. The product of a familiar debate within radical circles on the relative merits and dangers of reform and revolution, the non-reformist reform is Gorz's (1967) attempt to cut a generative middle path between socialist strategies that sought incremental, surface change (derisively labeled reformist) and immediate, structural change (revolution). Gorz instead argued for the development of a strategy built upon the implementation of what he alternatively calls non-reformist reforms and revolutionary reforms. Though it may initially appear a simple portmanteau, Gorz's strategy is slightly more complex. A reformist reform, Gorz (1967, 7) writes, "is one which subordinates its objectives to the criteria of rationality and practicability of a given system and policy" and "rejects those objectives and demands . . . which are incompatible with the preservation of the system." The danger of reformism, then, is less that it unintentionally preserves the oppressive system in question than that it expressly does so. Any goal of a reformist reform, however progressive, must be subordinated to the boundaries of the system, or, to return to Rancièrian language, the police order. This is a charge, in one form or another, that I heard frequently during fieldwork. Movement factions get co-opted through clientism, with small reforms or other material goods acting as the lubricant of urban political machinery. A law is passed, an arrest is made, or a megaproject is stalled, and the manifestaciones dwindle as anger subsides and public attention moves on from the cause. For whatever immediate good it might do for some or even all, reformism preempts revolution, and steals its lifeblood. As the work of Bayat (2013), Duhau (2014), Mitchell (2003), and many others illustrates, however, even marginal gains can both be extremely meaningful to their beneficiaries and can accumulate with geometric force over time, whether or not such potential is predictable or even imaginable contemporaneously. To dismiss such gains out of hand is, as Fainstein (2010, 19) argues, to fall prey to the most tired strand of Marxist teleology, which "leaves opponents of capitalist inequality with little to do short of revolution." Gorz also makes analytical room for what he calls "a not necessarily reformist reform . . . one which is conceived not in terms of what is possible within the framework of a given system and administration, but in view of what should be made possible in terms of human needs and demands." The third category of reform is the non-reformist or revolutionary reform, which is, quite simply, a categorically anti-capitalist measure. An initial difficulty with this analytical framework is its implicit temporality, as even Gorz notes the impossibility of knowing beforehand to which category a given measure may belong, at least in many cases. This is further complicated by the notion of unintended effects, as when a reform meant to be system-stabilizing mistakenly performs the opposite function. PRI economic policies in the neoliberal era might foster a pertinent example, wherein reforms meant to produce stable economic growth and salvage the credibility of the PRI failed to do either, but instead created opportunities for a growing opposition. Still, at the level of revolutionary strategy as practiced by many of Mexico City's grassroots organizations and groups, the notion of non-reformist reforms is analytically useful, though it requires some refinement.

The arc of Mexico City's recent history, at least from the era of the Mexican Revolution through the present, demonstrates a complex mixture of reforms that muddle categorical distinctions and bend even strategic definitions, and a lived reality of social revolution that flies in the face of existing temporal or even programmatic qualifications. To understand this history, Gorz's strategy of non-reformist reforms requires at least two transformative revisions. First, while it may be practically impossible to qualify in advance the position of a given reform vis-à-vis its revolutionary potential, Mexico City's history abundantly illustrates that such qualification is not always necessary. It need not always be known what good or ill may come of a particular law, program, or policy, for example, for these name only patterned relationships always subject to change. Moreover, even when they are allowed to run their course relatively undisturbed, the results of such arrangements can be highly unpredictable. In addition to the PRI's fateful neoliberal turn noted above, Jefe Mancera's disastrous attempt to carefully maneuver the Corredor project through the LPC provides ample evidence of how reforms intended (at least in part) to preempt popular pushback against a particular hegemony can be used to precisely the opposite effect. Non-reformist reforms, indeed reforms of any kind, this evidence suggests, can only be attempted in the present, in near complete ignorance of whether their intended effects will be realized. The benchmark victories along the winding trail of urban change for Mexico City's grassroots, many of which have been highlighted in the foregoing chapters, have been fought for, struggled over, tested, tried, and won. Their realization has depended not on the perfection of theory, but on the painstaking and often heartbreaking work of political practice. To be sure, no small measure of reflection and strategic thought find a prominent place in their story as well, qualifying them as works of praxis. But in most of the most meaningful instances, change came because people—under threats of violence, in the face of sure defeat, at the risk of reputation or livelihood, against their own better judgement, and over and against innumerable other adversities—showed up and did the work. As even Gorz's schematic makes clear, non-reformist reforms, as transformations from which gains can be symbiotically realized across multiple spatio-temporal scales, cannot always be designed. But it would seem they can always be made.

In global perspective, Mexico City's grassroots has also been especially adept at building broad-based movements and coalitions, at least since the Revolution. Echeverría (2019, 161) argues that even the vaunted student movement should be seen in this light, that it "belonged to the city and was original to it," that "the city felt itself involved in what young people were doing; it perceived that there was some relationship, if perhaps not very clear or very precise, of deep affinity between its own dreams, desires, or even resentments and longings for revenge, and what young people were doing." Such efforts and practices, in success and in what passes for failure, suggest another point of revision regarding the setting of political horizons. For many (though not all) of my friends, acquaintances, neighbors, colleagues, mentors, and others in Mexico City, in other words, the myriad social ills against which the capital's grassroots opposition have inclined themselves are irreducible to capitalism, per se. Theorists of justice, urban and otherwise, have in recent years succeeded in making this recognition a near commonplace in the Anglophone academic world, or at least its critical sectors.[4] An analytical position on urbanism cum capitalism à la Lefebvre pushes Gorz's argument closer to being able to handle such challenges as the environmental agenda of the CPCCMDC, for example, but can do little analytically to specifically address the issues of sexual violence that continue to plague the city and the country, or the gender and reproductive rights issues only now approached in the language of the 2017 constitution. At its most useful, the idea of a non-reformist reform (in what might be called its generic use) must subscribe to this broader orientation, striving and assessing the efficacy of measures not solely against revolutionary potential vis-à-vis capitalism, but instead against an expanded field of social and political systems and ills. If this case is instructive, at least such systems as patriarchy, racism, heteronormativity, and authoritarianism should also be under consideration. The potential exists for negative externality effects of non-reformist reforms as considered from this vantage, of course, such that a housing reform might deplete political energy from an anti-capitalist movement. The opposite potential must also be recognized, however, such that enhanced recognition and monitoring of human rights abuses might underscore the ravages of capitalism and help fuel its demise. Regardless, decades of chilango activism attest that progressive urban change must be about more than capitalism as such. Broadening the qualificatory horizon of non-reformist reforms in this way is an essential step in developing an

improved framework for evaluating the merits of even the simplest incremental alterations, by placing them within a more holistic theory of revolutionary urban change.

A third lesson is perhaps the simplest, and likewise refers to the path of
urban change. Urban theory and urban realities often present the world in
similarly stark dichotomies: us or them, acquiescence or revolt, exploiter
or exploited. As in the dyadic theoretical trap of reform and revolution,
chilangos have routinely been presented with similarly deceptive choices
between bad options. Cast your vote in a sham election and be called naive
(or complicit), or sit it out and be called jaded (or complacent). Accept the
way things are and get exploited, or stand up and get cut down to size, or,
if you're lucky, become part of the problem. Burn it all down at once or
accept that it will never change. But, as the chapters of this book have demonstrated and as the many chilangos featured (and not featured) in these
stories have always known, have learned, or have chosen to believe, such
political imperatives rarely present all the options. There is still room, this
account suggests, for surprise in urban politics. The forced mayoral retreat
produced by anti-Corredor forces, the rewarding refusal to admit the Charter's demise, and the resurrection of the revolutionary pantheon embodied
by the purple-clad faithful of the 2016 women's march and routinely con-
jured in the chants of "Zapata Vive!" all testify to the potency of refusal, creativity, and hope. Under the right conditions and given significant numbers,
for instance, even a deeply compromised democracy can serve to check
the powerful without needing to be dismantled. Praxis in the urban age,
all three of these lessons suggest, should be based on broader, more inclusive, and more transgressive ways of thinking about contemporary urban
geography and should proceed through bolder, more insistent, and more
hopeful ways of being and behaving as product and productive of the folds
of this global urban fabric.

Mexican revolution moves at its own pace, and is accomplished in its
own time. It is often incremental, its advance can be halting, and it can
sometimes even don an institutional guise. With the city's constitution in
force and the national Cuarta Transformación now fully underway, the city
is entering a period of uncertainty much like that of a century before, when
its great Revolution lumbered through its protracted endgame. The adoption of the 1917 constitution was followed by nearly two decades of violence
and disorder as rebellious groups remained at war with the state, as political
leaders were assassinated or forced into resignation or exile, and as fierce
debates raged over how to realize or curtail the constitution's principles. The
city may find itself in a similar position over the next several decades, as

the most radical provisions of its new constitution are tried and tested. Will the provision for the revocation of mandate help forge a new relationship between political elites and everyday citizens? Will rights to reproductive health and sexual identity and practice soften the scourge of sexual and gender violence that plague even the county's most progressive urban haven? If the history of post-revolutionary Mexico City is any guide, it can be said with surety that nothing born of such hopes is reasonably given. The city is the terrain and result of political struggle. But those who will fight these battles in, with, for, and against el monstruo have certainly crafted a powerful new weapon.

Appendix

An Explanatory Note on Approach and Methods

Writing a book with two souls, one ethnographic and the other historico-geographical, presents certain challenges, even over and above the invaluable analytical accommodations of a cultural studies- and urban studies-inflected dialectical materialism retooled for the urban age.[1] Contemporary Mexico City also presents special challenges of its own, for reasons beyond reckoning. Fortunately, many brilliant researchers and writers trod this ground long before I set out to conduct the research that became this book, and their accounts have lent inspiration and guidance from the very first.[2] Such scholarship helped me avoid, negotiate, overcome, and/or recover from many of the intellectual and practical difficulties of fieldwork and the analytical traps and mistakes of presentation common to such work. I owe an inestimably larger debt on the same account to the friends and strangers who participated in this research in one form or another, largely though not exclusively in Mexico City. But while I had many guides, this book shows the city by the light not only of its own contradictions, but also by those evoked by my particular methodological elections. In this appendix, I will briefly elaborate the specific methods by which I arrived at the material presented and the methodological approach that binds them into a unified study of contemporary Mexico City.

As I explained in the introduction, this project began with the paradoxical absence of the right to the city where and when I had expected to find it. Or, rather, this absence provoked a reconceptualization of the early research on which this book is largely based. I was initially drawn to Mexico City by the politics and movement surrounding the Mexico City Charter for the Right to the City, which appeared to me to be highly sophisticated, impressively inclusive, and more Lefebvrian than other instantiations I had then seen (though most folks associated with the Charter avoid mentions of Lefebvre, I would come to learn, and largely see their ideas and work as having little to do with Lefebvre's). I made preliminary research trips in 2013 and 2014 to gather information and begin building a research network, partly through introductions provided by friends and colleagues in New York and New Jersey. The generosity of several of these contacts allowed me to secure two formal affiliations for extended fieldwork during the 2015–2016 academic year, one with the Instituto de Geografía at the Universidad Nacional Autónoma de México (UNAM) and the other with Habitat International Coalition-Latin America (HIC-AL). I had planned to follow a movement focused on the right to the city, but soon encountered the disappointing dearth of the politics I had prepared to investigate. Colleagues at HIC-AL and UNAM were extremely helpful as I sought to make sense of the moment I'd arrived in and the political significance so many of them insisted it held for both the country and the city. Paying attention to and learning all I could about those issues that seemed to be most important to the people and organizations whom I encountered through participant observation, interviews, and archival work and with whom I was in conversation at HIC-AL and UNAM, I began to reframe my study around the relationships between contemporary militancies and the city's historical and geographical development, following several cases (three of which would become Chapters 3, 4, and 5) in particular.

As I recalibrated my fieldwork foci, I found myself in an extremely fortunate position. My affiliations at HIC-AL and UNAM allowed me access to vast and highly varied archival collections, as well as to invaluable connections to civil society, organized social movements and the broader political grassroots, and the GDF. For most of my 2015–2016 fieldwork, I lived in a rented room on the second floor of a stately, historic home tucked away inside a large block on a shabby section of Avenida Revolución in the colonia of Tacubaya, just around the corner from the Tacubaya Metro station. The friends and acquaintances I was privileged to meet at this house, especially through roommates Alex and Ignacio, also yielded important connections and comrades. Extensive time in the worlds of press archives, news coverage and commentary, social media (especially the activist circles I became interested and/or involved in), and other virtual spaces would also provide important sources of information. To organize the collection and analysis of these data, I employed specific sets of methods pertaining to each of three major data collection agendas.

The first of these methodological sets revolved around ethnographic practices, especially participant and passive observation and semi-structured interviews. Over the course of my fieldwork (roughly nine months over 2015 and 2016) and during five subsequent research trips between 2017 and 2020 (totaling just over three months), I recorded observations in a range of civil society spaces to which I was granted access, including formal meetings with GDF officials, formal meetings of the group working on the right to the city plataforma and other initiatives, formal or semi-formal strategic, organizing, and informational meetings (especially at the HIC-AL offices, or elsewhere at the suggestion or invitation of HIC-AL personnel or affiliates), and, on a semi-regular basis, quotidian interactions at the HIC-AL offices. In these meetings, I would sometimes participate in discussions, but would often only observe and record, switching from participant to passive observation as I felt comfortable and found appropriate (sometimes with the subtle and much-appreciated guidance of trusted colleagues).[3] I also conducted participant and passive observation in numerous public or semi-public fora across the city (plazas, museums, universities, GDF or federal buildings, and public streets, sidewalks, and parks), including during several manifestaciónes, such as the women's march described in Chapter 5. I employed a similar strategy in many social situations outside of formal civil society, especially as my social circles expanded and I found myself increasingly brought along, invited to, or crashing events in Tacubaya, Escandón, San Miguel Chapultepec, Polanco, Roma, Condesa, Cuahtémoc, Juárez, San Raphael, Doctores, Centro, San Ángel, and Coyoacán, among other places in and around the city.[4] These different spaces and the different groups they contained afforded me distinct vantages from which to approach and understand the city, and provided invaluable data in the form of fieldnotes. I nearly always carried at least one small notebook and pen, and jotted down sparse notes, which I later expanded upon, typed, and securely stored when I arrived home or to my workspace (at HIC-AL) or office (at UNAM).[5] Nearly all of the names of persons in these fieldnotes have been given pseudonyms, excepting cases of intentionally public behavior wherein a person identified themselves (as in a speech or other presentation) and public officials or other persons acting in an official capacity. Across these experiences, I have made sincere efforts to retain the mistakes, miscues, misunderstandings, gaffs (etc.) of my accounts, such that passages included in this book show to a reasonable degree the necessarily partial and refracted qualities of the scenarios described.

This set of methods also includes semi-structured interviews, most (though not all) of which were conducted during my initial fieldwork period in 2015 and 2016. In all, I conducted seventeen formal interviews with thirteen distinct participants, which spanned from roughly thirty minutes to over two hours in length. These are best considered expert interviews, as each of the participants was an expert of one kind or another (including government officials, NGO or civil society organization leaders, journalists, academics, industry experts [finance and hospitality], and business owners). These interviews were digitally recorded and later selectively transcribed and translated, sometimes with assistance from native speakers and/or the interviewees themselves. I also had many personal conversations that were, for a variety of reasons, not considered formal interviews, and were not audio-recorded, though most of these were recorded in my fieldnotes. Conversations ranging from decidedly casual to quite formal in bars, nightclubs, restaurants, university campuses, NGO spaces, museums, malls, parks, plazas, Ubers and taxis, and on canal boats, Metro trains, Metrobuses, and peseros and camiones also made their way into my fieldnotes in substantial numbers. Though the data from ethnographic fieldnotes and interviews exerted some influence on the production of the whole book, they were given more analytical weight in Chapters 4 and 5 in particular.

The second set of data collection methods pertains to archival data and historical analysis. My affiliation with HIC-AL was particularly important to my efforts in this area, especially during fieldwork in 2015 and 2016. HIC-AL's library contains a wealth of information, analysis, finished reports, and much else from the organization's extensive history of research and advocacy in the city, along with that of their many partners. From this archive, I was able to view and make copies of annual reports, research reports prepared by HIC-AL and affiliates, theses written by affiliated students, and essays on a wide array of topics, among other useful documents. Newspaper archives also proved a significant source of information, with (largely online) access variously provisioned by university affiliations before, during, and after fieldwork periods.[6] Though other titles were surveyed, the bulk of the material I consulted came from Mexico City dailies *La Jornada* (based at UNAM), *El Universal, Excélsior,* national daily *El Informador;* and international outlets with a significant presence in the city, including *Reuters, El País,* the *Guardian,* and the *New York Times.* I also collected extensive material from the reporting and commentary found in political and economic magazines, e-zines, and other such publications, including *Proceso, Animal Político, Nexos, Sin Embargo,* and *El Economista.* I also gathered cartographic, demographic, and electoral information from a variety of online archives and repositories, most notably INEGI, INE, IFE, and IEDF, and from secondary sources. A last archival data collection strategy involved scouring for government reports and/or data sets, official and commercially produced guidebooks and promotional materials, rare books/newspapers/magazines, and other published ephemera at several libraries (notably at UNAM and Princeton University) and more especially the matchless bounty of the capital's secondhand bookstores and junk shops, particularly in and around Centro, Tepito, and Coyoacán. I sorted and organized these materials both temporally and thematically throughout the process of data collection, which extended through final editing of the initial draft of this book in early 2021. Such archival data inform the whole of the book, but are afforded particular analytical weight in Chapters 1 and 2.

I deployed yet a third set of methods to collect a variety of data in virtual space, especially during my fieldwork. I was advised early on that Facebook and more especially Twitter (and later WhatsApp) have become important organizing and promotional tools for grassroots politics in Mexico City, much as they have in other urban contexts, and I closely followed as large a number of groups, platforms, discussions, and influential individuals as practical.[7] While these platforms initially provided me with information about events and meetings, I soon began to monitor them as another valuable source of supplementary data. While using social media, I acted largely as a passive observer, sometimes taking notes on and sometimes digital copies of interactions, promotional materials, or commentary. Over time, these ballooned into a rather large repository of notes and images, which I likewise sorted and organized thematically. As my initial fieldwork progressed, and especially in subsequent rounds of fieldwork, I became a more active participant in some discussions, occasionally posting photos or descriptions of events I had attended or planned to attend, or shared other newsworthy events I had witnessed (such as the grossly underreported collapse of the historic Cine Ermita just a block from my home on Avenida Revolución). I also watched uncounted hours of video, mostly on YouTube and Vimeo, sometimes with friends or colleagues. The content of these videos (which were sometimes sent to me directly, but more often linked or posted on social media, especially Twitter) included political ads and promotional content, speeches, interviews, stories, commercials, guerilla reporting, analysis, marches, rants, arguments, travel logs, walking tours, clips from TV shows, and much else. In some cases, I transcribed and translated portions of these videos, made screen captures, or took notes, organizing these thematically along with other digital materials. I also collected and reviewed a great deal of news coverage, often via links posted on social media platforms, and also via six tailored Google alerts I set up and monitored for over four years, beginning in 2014. Though these data inform the project at the level of general trends and the conversational tenor of particular topics (such as the Corredor project), and were sometimes mined for exemplary evidentiary material, I was careful to account for the many limitations of such data, not least the limited access many chilangos have to such platforms and the unevenly distributed understanding of their use and utility across the city.

Abbreviations

ALDF. Asamblea Legislativa del Distrito Federal (Legislative Assembly of the Federal District)

CCMDC. Carta de la Ciudad de México por el Derecho a la Ciudad (Mexico City Charter for the Right to the City)

CDHDF. Comisión de Derechos Humanos del Distrito Federal (Human Rights Commission of the Federal District)

CPCCMDC. Comité Promotor de la Carta de la Ciudad de México por el Derecho a la Ciudad (Promotional Committee of the Mexico City Charter for the Right to the City)

ECLAC (CEPAL). UN Economic Commission for Latin America and the Caribbean (Comisión Económica para América Latina)

GDF. Gobierno del Distrito Federal (Government of the Federal District)

HIC-AL. Habit International Coalition-América Latin (Habitat International Coalition-Latin America)

IECM. Instituto Electoral de la Ciudad de México (Electoral Institute of Mexico City)

IEDF. Instituto Electoral del Distrito Federal (Electoral Institute of the Federal District)

IFE. Instituto Electoral Federal (Federal Electoral Institute)

INE. Instituto Nacional Electoral (National Electoral Institute)

INEGI. Instituto Nacional de Estadística y Geografía (National Institute of Statistics and Geography)

LPC. Ley de Participación Ciudadana (Law of Citizen Participation)

MORENA (OR MORENA, OR MORENA). Movimiento Regeneración Nacional (National Regeneration Movement)

PRD. Partido de la Revolución Democrática (Party of the Democratic Revolution)

PRI. Partido Revolucionario Institucional (Institutional Revolutionary Party)

PAN. Partido Acción Nacional (National Action Party)

PARM. Partido Auténtico de la Revolución Mexicana (Authentic Party of the Mexican Revolution)

UNAM. Universidad Nacional Autónoma de México (National Autonomous University of Mexico)

Notes

INTRODUCTION

1. *Jefe de gobierno*, "head of government," usually translated "mayor," is the offi-
    cial title of the city's chief executive.
2. *Plataforma* is often used to describe what might in English be called "move-
    ments" or "social movements," as the Spanish *movimiento* has taken on a more
    specific meaning, discussed in more detail in later chapters.
3. As Holston (2008, 33) instructively argues in their similarly structured eth-
    nographic study of an urbanizing São Paulo within the history and geogra-
    phy of a modernizing Brazil, "This history is not past. It continues to struc-
    ture the present."
4. In this book as elsewhere, urban revolution is a complex concept with a variety
    of complementary and competing determinants, meanings, and uses, largely
    though far from exclusively rooted in the work of Henri Lefebvre (2003a).
    Many of these are explored in detail in Chapter 5, especially as they appear
    within the four vectors I trace there through the revolutionary structure of feel-
    ing I argue pervades contemporary Mexico City. See especially Smith (2003).
    See also Brenner and Schmid (2014), Merrifield (2014), Robinson (2016),
    Boudreau (2017, especially pages 7–18), Murray (2017), Beauregard (2018, es-
    pecially pages 15–20; 152–54), and Martínez et al. (2020).
5. Paramount among the many points of reference for my usage of *leviathan*
    is the work of Davis, especially their *Urban Leviathan: Mexico City in the
    Twentieth Century* (1994), which is perhaps the most widely read and cel-
    ebrated Anglophone book on Mexico City of the last half-century, at least
    in urban studies, and serves as a major source of inspiration for the present
    work. Following Davis, leviathan is used throughout this book (especially in
    Chapter 1) to refer to both the PRI and PRI–state (primarily the latter) and to
    the city itself (especially since the middle decades of the twentieth century),
    both of which are commonly garbed in monstrosity by residents and visitors
    alike (to say nothing of analysts, among whom such descriptors are also notice-
    ably common). *Urban Leviathan* was and is pioneering in its insistence on the

complex interplay between scales of authority in Mexico City, and also for its spatial sensitivity, both of which are important axioms for the present work as well. Much has changed in Mexico City since the publication of that seminal text, and much has also remained the same; a reality on which this book seeks to build. Importantly, *Monstrous Politics* is centrally concerned with grassroots politics in the context of an ongoing urban revolution, most especially with efforts to stake specific—but nevertheless extremely complex—kinds of political claims. Within the century or so covered in this book, such demands have been made of both the political leviathan (the PRI–state) and of various guises and iterations of its urban counterpart (including the municipal state, the material and symbolic spaces of the city, and its various neighborhoods and communities). In my analysis, as for many of the friends and strangers I spoke with, read about, labored alongside, and observed while conducting the research for this book, the two are profoundly and inescapably linked, much as I understand them to be for Davis.

6. *Chilango* is a common nickname for residents of Mexico City. *Chilangolandia* is therefore "land of the chilangos." Some of the many conflicting accounts of the name's genesis have it that it once referred solely to newcomers to the city, as opposed to "native" residents of longer tenure. In common parlance throughout the city, however, the name can now largely be understood to refer to any resident of the capital, and indeed the urban region. The less popular *defeño* no longer seems an apt characterization, as it names a resident of DF, the Distrito Federal (Federal District), which was legally renamed la Ciudad de México (Mexico City) in January of 2016, as discussed in later chapters.

7. My understanding of and approach to dialectics also owes much to the teaching of both of Don Mitchell and David Harvey, and to the patience of Rick Schroeder.

8. Lest it be assumed that I chose this framework arbitrarily, or that this choice is the product of a blind adherence to Marxian dogma. Indeed, my arguments about dialectics in Chapter 3 are likely to ruffle some liturgical Marxist feathers.

9. This inconstancy and malleability are differently expressed, for example, in the epigraphs to this introduction, by vaunted twentieth-century author Octavio Paz and locally revered *Tepiteña* Lourdes Ruiz Baltazar, the latter better known as *la Reina del Albur* (Queen of [the] Albur).

10. See Jackson et al. (1979); Mitchell (2008).

11. See, for instance, Tenorio-Trillo (2020, 71), whose experience of walking in the city and its evocation of the 1985 earthquake and its 2017 counterpart illustrate this profoundly affective dimension of urban space: "But for my generation, to walk the city after an earthquake is to converse with the city in tones of postcoital lucidity and resignation. How is it possible that another September 19th, 32 years later, has brought me back to my essence? I walk and walk, once again reduced to the terror of living and being this 'I' that I was before, which I have never stopped being."

12. Owing to its rather exceptional position at the center of Mexican political economy, Rodríguez Kuri (2012, 17) argues that Mexico City is subject to a unique "geopolitical overdetermination." This characterization places analytical emphasis on the role of (inter-scalar) conflict in the development of the city's political history, a posture I likewise adopt in examining the changing morphologies of city's political geographies.

13. Roughly, "we are, because we are land," González's words highlight not only the intimate connections between people and place (city, neighborhood, community) and people and land (earth, soil, ground) but also, as he has explained, that political participation and belonging, the rights and duties of citizenship, and the integrity of territorial political communities are fundamentally conditioned by access to and presence in the spaces of the city. See also Tamayo (2006, 29), for whom "la cohesión de los ciudadanos o su diferenciación se da por la cultura, la cual implica permanencia, pertenencia y, sobre todo, prácticas, y por lo tanto el 'estar ahí' en un territorio" (the cohesion of citizens or their differentiation is given by culture, that which implies permanency, belonging and, above all, experience, and therefore the "being there" in a territory).

CHAPTER 1

1. See the United Nations' report *World Urbanization Prospects* (2018 revision), which lists Mexico City as the world's fourth largest city, behind only Tokyo, Delhi, and Shanghai; 2020 projections, however, place Mexico City in fifth position behind, respectively, Tokyo, Delhi, Shanghai, and São Paulo.

2. A label often incorrectly attributed to Alexander von Humboldt (see de Mauleón 2015), the name *la ciudad de los palacios* was given to the city by the long-winded English traveler—and future governor of the English Colony of Victoria, Australia—Charles La Trobe, whose travel diary, *The Rambler in Mexico*, describes his entry into Mexico City thus (La Trobe 1834, 102): "And, when approaching the main valley, the villages thicken around him, with their streets, cheered and beautiful amid the general sterility, by groups of the graceful peruvian [*sic*] pepper-tree; and the roads are seen crowded by long strings of laden mules, and gay cavaliers,—and the stupendous works of human design, harmonize with those of nature, and prepare him for the sight of one of the most extraordinary scenes in the world, whether we regard the works of men, or those of God, the Artificer of all. And such is the Valley and City of Mexico."

3. See Kandell (1988), Davis (1994), Bruhn (1997), Monsiváis (1997), Ibargüenguitia (2004), Lida (2008), Ross (2009), Hernandez (2011), Tenorio-Trillo (2012), Vitz (2018), and Boudreau and de Alba (2020), among countless others. It would be a mistake, however, to reduce this monstrosity to any of its myriad caricatures. As Boudreau and de Alba (2020, 65–66, following Foucault) explain, the ubiquity of the city's monstrosity across academic and popular genres represents a "mode of political imagination" that perversely blends "the prohibited, the

impossible, biological errors, and violation of the law" into "a humanized, yet agonizing and sick image, based on the affective experience of its urban writers (metonym) and the symbolic reputation of the capital city within the country." Yet even their careful disaggregation borders on the binary (good/evil, chaos/order, planned/unplanned) in describing a city whose quotidian intimacies refuse to be so qualified, as indeed these authors and others have amply demonstrated. Monstrosity, in other words, is hardly contained by its repulsive dimensions. Mexico City maintains an impossible appeal both in spite of and, importantly, because of its horrors (see Tenorio-Trillo 2020).

4.   Though the city of course has its international apologists as well, seemingly increasing in the last decade or so. The *New York Times*, for instance, ranked it first among its "52 places to visit" in 2016. The city's growing international reputation as a hip urban scene will figure prominently in Chapter 4.

5.   See also Dagnino (2007) on the "perverse confluence" of citizenship in contemporary Latin America.

6.   See also Rodríguez Kuri (2012).

7.   I employ the language of *crisis* with due caution for its tendency (not least in comparative urban studies) to evoke chaos, corruption, degradation, (etc.), often conspicuously associating these with Global South contexts. Nevertheless, both in my observations and, more significantly, in the wide literature on Mexican political economy and endogenous reporting on politics, economics, and many other topics, invocations of crisis are a commonplace bordering on the banal. Indeed, the recurrence of the concept across narrative genres in political discourse and common parlance is a primary reason for my usage throughout this book (to be explored more fully in Chapter 3).

8.   See Gillingham and Smith (2014).

9.   My delimitation of this period owes much to the work of, among others, Tenorio-Trillo (2012, xv), who argues that it was in the period between 1880–1930 that "Mexico City started upon the route toward what it is today, namely, a megalopolis, an ecological disaster, and the enchanting monstrous capital of a modern nation." As will become clear, similar arguments could also easily be made about the relative historical and geographical gravity of both of the other two periods considered in this chapter, though on altogether different footing.

10.   *Caciquismo* and its roles in Mexican history and historiography (as well as the political present) enjoy a robust literature; see Knight and Pansters (2006). The word cacique is of Indigenous origin, but in contemporary usage is typically translated "boss."

11.   *Caudillo* can be translated as "strongman," "leader," "figurehead," or "warlord," and refers to both to a person and to a style of leadership typically characterized by a charismatic personality at the head of a loyal militancy (or military force). See Williams (2011) for a review of the shift in such a "charismatic nature of authority" (sometimes called *caudillismo*) from person to position in the second half of the twentieth century in Mexico.

12. The *New York Times* stated in 1910 that Mexico's railroads had grown from 578 kilometers at the beginning of Díaz's reign in 1877 to 24,160 kilometers as of 1909.

13. See Neufeld (2012) for a more nuanced account focused on the unruly Presidential Guard during the waning years of the regime. See also Beezley (2014).

14. "Progressive Mexicans," Lear (2001, 33) argues, "were particularly sensitive to the perceptions of foreigners, and the various national delegations that were to attend the centennial provided an opportunity for Mexico to assert its claim to membership among modern nations." See also Tenorio-Trillo (1996) on the attitudes of these cosmopolitan *capitalinos.*

15. Joseph and Buchenau (2013, 36) claim that Madero even enjoyed "the tacit support of the US state and business community" at the outset of the Revolution.

16. This tour is said to have created an abiding affection for Cárdenas—who soon took on the nickname *Tata Lázaro* (Papa Lázaro)—that would remain strong enough to enhance popular support for his son Cuauhtémoc as a presidential candidate nearly sixty years later.

17. See Joseph and Buchenau (2013).

18. A *sexenio* is a six-year term. The constitutional change from four-to-six-year terms was made in 1928 under Calles.

19. Article 27 of the 1917 constitution defined such ejidal grants, which Cárdenas used to greater effect than perhaps any president before or since. Hanson (1971, 32–34) reports that "by 1940 virtually one-half of Mexico's rural population lived on *ejidos,*" and that Cárdenas's grants benefitted 811,157 recipients, nearly doubling the total number of recipients since 1920 and more than double the number of the next closest PRI administration through 1964 (Ernesto López Mateos, whose grants went to 304,498 recipients). See also Hernández Chávez (2006) on Cárdenas's expropriations and land distributions. Rapid urban expansion would soon engulf many of the ejidos in and around the capital, making them attractive sites for informal housing construction. Varley (1987, 469) reports that "by 1970, at least 16 per cent of the Mexico City population was resident on *ejido* lands, which constituted 20 per cent of the urban area." Arguably the most important piece of Cárdenas's nationalization program came in 1938, when Cárdenas nationalized Mexico's oil industry. Though a decidedly unpopular move abroad (especially in those countries, like the United Kingdom, whose investors had interests in Mexican oil), it remains one of the pillars of Cárdenas's highly durable legacy.

20. As Rodríguez Kuri (2010) demonstrates, revocation of the Federal District's sovereignty had been one of two radical proposals regarding the capital proposed to the Constitutional Congress of 1916–1917 by Revolutionary and President Venustiano Carranza, the other being a massive expansion of the capital's boundaries to include the entirety of the Valley of Mexico. Though neither proposal was adopted in the Constitution of 1917, the former became a

reality just over a decade later (the latter, which has been proposed many times since, continues to be an unrealized dream for many, as discussed in Chapter 5 in particular). Ward (1989) argues that this 1928 move came at the behest of soon-to-be assassinated Obregón, and was aimed at bringing political squabbling among the municipios under control. Davis (1994, 166) likewise explains that "national political leaders saw the elimination of democratic practices as the only way to control labor politics and urban development in the capital."

21. *Municipios* are the governmental subunit of all Mexican states. *Delegación* was a special governmental subunit for the Federal District, typically rendered borough in English.

22. *Regente* is usually translated into English as mayor; this position was formally that of *regente* (regent) until 1997, and *jefe de gobierno* (head of government) thereafter.

23. Such arrangements are often labeled *presidencialismo* (presidentialism). See Davis (1994) for an insightful discussion of the role presidentialism has played (and arguably continues to play) in scholarship on Mexico. See also Perló Cohen (1993).

24. Here it will be useful to distinguish what is sometimes called the metropolitan area or Greater Mexico City from the Federal District (and later Mexico City), a distinction that will become increasingly important through the middle decades of the twentieth century. The metropolitan area—variously referred to as the Área Metropolitana (Metropolitan Area), the Valle de México (Valley of Mexico), the Zona Metropolitana del Valle de México (Valley of Mexico Metropolitan Zone, hereafter ZMVM), and especially the Zona Metropolitana de la Ciudad de México (Mexico City Metropolitan Zone, hereafter ZMCM)—also includes several municipios of the neighboring States of México and Hidalgo, depending of course on which term is used, and, in some cases, on who is using the term (see Connelly [2003] for clarification). According to the Instituto Nacional de Estadística y Geografía (National Institute of Statistics and Geography, hereafter INEGI), the ZMCM includes one municipio of the State of Hidalgo and fifty-eight municipios of the State of México (see Figure 5.9). In the most expansive contemporary demarcation built around INEGI data (ZMVM), Ziccardi (2014) sets the 2010 population at 21 million, inclusive of the Federal District's sixteen delegaciones, fifty-nine municipios in the State of México, and twenty-one municipios in the State of Hidalgo. Figures for the period under examination here also vary with terminology and usage. Davis (1994) reports the Federal District's population at 1.645 million as of 1940, and that of the Mexico City Metropolitan Area at 1.758 million. Making no distinction, Gilbert and Ward (1982) report the population of what they call Mexico City at 1.8 million in 1940. Davis (1994) also notes a third category in the national census data, that of Mexico City (presumably different from that employed by Gilbert and Ward, who likely refer to the Metropolitan Area, given the figure they posit), which is there reserved for only the inner areas of the old city, specifically the delegaciones of Miguel Hidalgo, Cuauhtémoc,

Benito Juárez, and Venustiano Carranza, an area Ward (1990, 18) argues is also loosely affiliated with "occasional reference to the 'inner-city' or downtown area." Ward (1990, 17–18) concisely and correctly explains that despite a plethora of qualified attempts, "the definition of what constitutes Mexico City is not fixed." For the remainder of this chapter, *Mexico City*, unless otherwise specified, will be understood to include the areas commonly associated with the metropolitan area (following Connelly's (2003) usage). The issue was further complicated by the reestablishment of the legal entity Mexico City (la Ciudad de México), which replaced the Federal District as of January, 2016.

25. Much of the Templo Mayor would not begin to be fully excavated until the 1970s.

26. Planned as the vast Federal Legislative Palace, the project sat unfinished for roughly twenty-five years until the cupola that would have formed its center was repurposed as a standalone Monument to the Revolution in 1938.

27. See also Piccato (2001).

28. *Colonia* (literally, colony) is now an official designation conferred by the city government. As of 2010, the Electoral Institute of the Federal District (IEDF) counted 3,480 colonias across the District's 16 delegaciones.

29. Most economic and political texts place the break point for this history at around 1970, for two reasons. First, arguably the most significant event for the PRI came in 1968, when its forces murdered a large number of protesting students in what came to be known as the Tlatelolco Massacre (discussed below) which many see as a turning point for the party's relationship with the middle classes, with the public, and the city. Second, the years of the Miracle were largely (though not entirely) over by 1970, when economic growth began to stagnate and the contradictions and shortcomings of ISI policy (discussed below), and perhaps even more importantly widespread corruption and graft, hurt the party's ability to deliver on its promises. While both arguments are compelling, the aim of this chapter is to elaborate an analysis that can inform, properly contextualize, and find resonance with larger arguments about the city and its politics more fully articulated in later chapters, rather than focusing on national economic policy and trends or the political crises of the PRI, per se (work that has been undertaken extensively by others). Though both of these figure prominently in this chapter, I selected the break point of 1980 (as opposed to the more common 1970, or even 1982, which would also be compelling both as a beginning of a new presidential sexenio and the onset of a significant debt crisis) for its more compelling illustration of the city's population and spatial growth, both of which remained rapid through the 1970s even as the national economy began, however haltingly, to descend from the heights of the previous three decades. See Knight (1990) for a brief critical review of historiography of this period.

30. This airport was halted for the second time by then-President Elect Andrés Manuel López Obrador after a controversial referendum in 2018, some three weeks before taking office. López Obrador (hereafter AMLO) subsequently

announced alternative plans to construct an international airport on a nearby military base (Santa Lucía), which he claimed would resolve the chronic service issues of Benito Juárez International Airport for the next "40 or 50 years" (Méndez 2018).

31. See Harth Deneke (1966), Perló Cohen (1979), Duhau (2014), Guttman (2002), and Ward (1990). For regional perspective, see also Holston (2008) and Goldstein (2004) on urban informal settlement in Brazil and Colombia, respectively.

32. See Perlman (1975). See also Wacquant (2008) on advanced marginality.

33. See Ward (1978, 48), for instance, who found that while "squatter settlements by themselves do not provide a means for upward socio-economic mobility," a range of potential opportunities for the improvement of life chances did exist in the Mexico City squatter settlements of this period, depending on the varying quality and perhaps durability of the connections residents are able to make to "the wider economy and social structure," such as "educational facilities, productive well paid employment and so on." See also Varley (1987).

34. Such practices bear strong resemblance to those documented in a variety of urban contexts during and after this period, and are a hallmark of what Caldeira (2017) calls "peripheral urbanization."

35. On urban informality, see also Rakowski (1994); Roy and Alsayyad (2004); Fischer, McCann, and Auyero (2014).

36. It would be misleading, however, to gloss over the pockets of extreme density that continued to exist in and around the historic center and in neighboring areas of the old city core, such as the infamous Tepito, long known for its black markets, second hand or counterfeit goods, crime, and more recently for one of the most visited and longstanding shrines to Santa Muerte (the apocryphal saint known as Holy Death). The most common type of housing in such areas is the *vecindad*, a tenement style structure in which families typically inhabit one windowless room and share common toilets and other basic facilities. Surveying those of the historic center, Eckstein (1988, 45) explains that "vecindades are densely populated, unsanitary, and poorly ventilated. Plumbing facilities frequently do not function properly, and buildings at times are so run down that sections occasionally collapse, leaving survivors injured and homeless." Portraits of life in such conditions can also be found in Lewis's (1959) *Five Families: Mexican Case Studies in the Culture of Poverty* and *The Children of Sánchez: Autobiography of a Mexican Family* (1961), the latter of which is set entirely in Tepito. Lewis (1961, xiii) describes Tepito as "a poor area with few small factories and warehouses, public baths, run-down third-class movie theaters, over-crowded schools, saloons, *pulquerías* [taverns where *pulque*, a native alcoholic drink, is sold], and many small shops." As for the vecindades themselves, Lewis (1961, xiv) provides this introduction to that occupied by the Sánchez family: "Spread out over an entire square block and housing seven hundred people, the Casa Grande is a little world of its own, enclosed by high

cement walls on the north and south and by rows of shops on the other two sides. These shops—food stores, a dry cleaner, a glazier, a carpenter, a beauty parlor, together with the neighborhood market and public baths—supply the basic needs of the vecindad, so that many of the tenants seldom leave the immediate neighborhood and are almost strangers to the rest of Mexico City. . . . Two narrow, inconspicuous entrances, each with a high gate, open during the day but locked every night at ten o'clock, lead into the vecindad on the east and west sides. Anyone coming or going after hours must ring for the janitor and pay to have the gate opened. . . . Within the vecindad stretch four long, concrete-paved patios or courtyards, about fifteen feet wide. Opening on to the courtyards at regular intervals of about twelve feet, are 157 one-room window-less apartments, each with a barn-red door. In the daytime, besides most of the doors, stand rough wooden ladders leading to low flat roofs over the kitchen portion of each apartment. These roofs serve many uses and are crowded with lines of laundry, chicken coops, dovecotes, pots of flowers or medicinal herbs, tanks of gas for cooking, and occasional TV antenna [sic]."

37. The mammoth, three-section Bosque de Chapultepec (literally Chapultepec Forest), which contains Castillo Chapultepec (Chapultepec Castle), the imperial and later presidential residence until the sexenio of Lázaro Cárdenas, who moved to the symbolically less-palatial residence known as Los Pinos (which remained the presidential residence until the 2018 election of Andrés Manuel López Obrador), also within the park. Bosque de Chapultepec is one of the world's largest urban parks, at 1,655 acres. The name Chapultepec is derived from the Nahuatl word meaning grasshopper.

In the common lexical cultural geography of contemporary Mexico City, Lomas is a bit of a loaded, catch-all term, but should be understood to include at least the eight sections of Lomas de Chapultepec in the delegacion of Miguel Hidalgo, and the areas known as Lomas de Reforma and Bosques de las Lomas. At the very least, this area is known for its ostentatious wealth and a certain social, economic, and especially legal insularity. For instance, Lomas is where my roommate and close friend Ignacio reportedly attended a party that lasted several days, to which some of the guests had arrived by private helicopter to a helipad on the grounds of the home where the party was held. Incidentally, after making passing reference to the quality of the knife set he saw while wandering through a kitchen near the end of his stay, he was invited by the host to take home the whole set. He accepted them gladly, but reportedly forgot them in the car they sent him home in or otherwise lost them in transit some hours later (fieldnotes, February 20, 2016). These are the sorts of parties that have been referred to by friends of mine as tiger parties, as they are thought to be thrown by people who might actually own pet tigers (friends also told me of offers to meet this or that person's pet tiger). These are the kinds of references made to the ostentatious wealth of the area, and they form part of the general attitude toward this part of the city, at least among those who don't reside there.

An introduction to a similar portrait of the disconnectedness commonly associated with the wealthy residents of the area is perhaps best provided by the recurring sketch "Las Chicas VIP" on the comedic television series *Desde Gayola* (Villalobos 2002–2013), most especially the episode in which an upper class girl caps her condescending explanation of her hesitance to formalize her relationship with her *naco* (a racist and classist pejorative) boyfriend with what she views as the self-evident contradiction of his residence in Colonia Navarte (a middle-class neighborhood south of the city center) and hers in Lomas. See also the trenchant portrayals offered by Blanco (2004) and the photography of Rossell (2004).

38. See Story (1986) for a discussion of the Mexicanization policy of requiring majority Mexican ownership first enforced in 1944 to curb the influence of foreign capital and what was perceived as dependence on foreign developed economies, pursued under successive PRI administrations during the Miracle, especially notable in the production of steel, cement, glass, fertilizer, cellulose, and aluminum, among other inputs for building and construction and commodity production. See also Martínez (2016). ISI policy was a popular program in Mexico during the Miracle decades, as it was throughout the region, most influentially developed as a comprehensive theory by Raúl Prebisch, second Executive Secretary of the UN Economic Commission for Latin America and the Caribbean (hereafter ECLAC, known in Spanish as CEPAL), and heavily influenced by dependency theory. Dependency theory of course also played a major role in Latin America during the twentieth century, not least in appraisals of Mexico. See Cardoso and Faletto (1979).

39. There are, of course, important caveats to these general national trends, including especially the establishment of the industrialization program for the US-Mexico border region resulting in incredible growth of the maquiladora manufacturing industry beginning in the early 1960s. Nacional Financiera was created under Cárdenas in 1934 but took on increasing importance during this period.

40. As Moreno-Brid and Ros (2009, 93) explain, "For the next thirty years, Mexico's economy grew at a sustained annual pace of 6.4% in real terms and per capita gross domestic product (GDP) grew at a rate of 3.2% per year (INEGI, 1999a). Manufacturing was the engine of growth, with rates of growth of production of 8.2% per year (INEGI, 1999a) and, for most of the period, the dynamic domestic market was its major source of demand. The country was transformed from an agrarian society into an urban, semi-industrial one."

41. The Confederación Nacional de Organizaciones Populares (National Confederation of Popular Organizations, hereafter CNOP) was founded in 1942, and while its structure was extremely broadly articulated, it largely represented the interests of the growing middle-class and state sector employees. See Davis (1994); Alexander (2016).

42. See Eckstein (1988, 19).

43. See the detailed analysis of Thornton (2021, especially pages 145–165), who argues that Mexico's priorities and policies on the global economic stage shifted considerably during the Miracle decades as foreign investment and debt ballooned to staggering levels by 1970.

44. See del Carmen Moreno Carranco's (2008) discussion of the development of the *ejes viales* (downtown thoroughfares constructed through aggressive street-widening projects and property expropriations). Known colloquially as the professor, Hank González and his sons Carlos Hank Rhon and Jorge Hank Rhon have been widely accused of various financial, political, and narcotics-related crimes over the last several decades (see Bergman n.d.; Farah 1999; *New York Times* 2001). The elder Hank was a powerful political operator for several decades in the PRI, and, as former governor of the Estado de México and prominent figure in the PRI, was influential (through his prominent role in the shadowy Grupo Atlacomulco) in the party's return to presidential power in 2012, in the person of Enrique Peña Nieto, then the outgoing governor of the Estado. The patriarch of one of Mexico's wealthiest families, Hank was a tragic embodiment of the phrase *political economy*, and was widely known for his summation of Mexican politics in the infamous statement "a politician who is poor is a poor politician" (Tuckman 2001).

45. Joseph and Buchenau (2013) make a similar argument about the party's hold on power as far back as its earliest years, as they claim the party was always required to make concessions and to work to a large extent within the confines of existing political realities across the country's many regions.

46. Ward (1990, 33) states, "During the early decades of the city's growth, when the demand from industry for labor was high, migration flows accounted for around 60 per cent of population expansion, with the remainder the result of natural increase (Unikel 1972). But, in the absence of a sharp decline in the birth rate achieved nationally or locally, natural increase quickly took over as the principal component of city growth." While empirically verifiable, these statements may serve to slightly misrepresent the dynamics of metropolitan growth. While natural increase indeed seems to have displaced rural-to-urban migration as proportionally the largest year-to-year factor during the early part of the Miracle period, Ward's discussion of the demographic character of migrants (especially young people likely to start new families relatively soon after arriving in the city) suggests that it plays a considerably larger ongoing role than a simple rate calculation can illustrate. This is simply to say that it is reasonable to expect the expanding peripheries of the city continued to be heavily influenced by the experiences of rural-to-urban migration for a longer period than crude figures may suggest.

47. Which has produced no small amount of contest over its memory among contemporary residents, alongside a certain predictable institutional forgetfulness. See Pensado (2013).

48. Near the outset of the extended addendum to his *Labyrinth of Solitude*, Octavio Paz (1985, 231–32, original emphasis) describes the development of this student movement thus:

> The student movement began as a street brawl between rival groups of adolescents. Police brutality united them. Later, as the repression became more severe and the hostility of the press, radio, and television—almost all pro-government—increased, the movement strengthened, expanded, and grew aware of itself. In the course of a few weeks it became clear that the young students, without having expressly intended it, were the spokesmen of the people. Let me emphasize that they were not the spokesmen of this or that class but of the collective conscience. . . .
>
> The students' demands were genuinely moderate: derogation of one article of the Penal Code, an article that is completely unconstitutional and that contains the affront to human rights called "crime of opinion"; the freeing of various political prisoners; the dismissal of the chief of police; et cetera. All of their petitions could be summed up in a single word that was both the crux of the movement and the key to its magnetic influence on the conscience of the people: democratization.

In contrast, Bruhn (2008, 120–21) states that the movement

> started with a march in support of the Cuban Revolution by high school and college students in July 1968. Police entered the campus of several high schools and the National Autonomous University to break up the demonstrations. Faculty and students considered the police actions a violation of the university's legal autonomy. They organized a series of protests, adding to their original complaints about price repression a growing list of demands for democratization, better living conditions for the urban poor, and postgraduation [*sic*] career opportunities. The government—usually tolerant of student marches—grew concerned about the increasing size, aggressiveness, and external support for student protests.

Accounts roughly converge on the movement's broad and representative character, though there are differences of emphasis placed on the origins of the movement's concerns and the relative weight they receive in the movement's lasting meanings. Gutmann (2002, 71) argues, "Objecting to Mexico's assertion that it was a modern, democratic society, and to the claim that Mexico was no longer really a third-rate power, the students of 1968 were determined to reveal the realities of poverty and misery and corruption in their country." In a similar vein, Crane (2015, 4) sees the student movement as arising from several decades of activism among such groups as urban industrial workers and women seeking the national franchise, conditions "set by dispersed acts, demands, and justice claims in the decades before." Taking a firmer position, Eckstein (1988, 218) places distance between the student movement protestors of 1968 and those seeking more radical changes in the realm of political economy, such as those in Mexico City who openly sympathized with Cuban Revolutionaries or the Chinese Communist Party of Mao Zedong: "Many of those who participated in the 1968 protests were not Marxists but critics of specific government policies. They focused on the lack of civil liberties, not economic inequalities."

49. The *dedazo* (roughly, the finger tap, or the big finger) is the process by which the PRI has historically chosen its presidential candidates. *Presidencialismo* refers to the tendency to hoard the powers of the state in the executive branch and the person of the president.

50. "In Mexico," Echeverría (2019, 166–67) writes, "young people detected a series of lies spun by the state that for many decades had exercised a kind of totalitarianism, a soft totalitarianism, if you will, but ultimately a totalitarianism, which insisted, nevertheless, on presenting itself as a democracy. This hypocrisy was unbearable for the youth, so they posed a challenge to the government: to recognize that it pertained to a totalitarian state, founded on violently repressing the unhappy oppressed, or to show that what it insisted so much on when it presented itself as a democracy was not true."

51. The Plaza of Three Cultures is so named for its architectural environs, comprising the ruins of a pre-columbian Aztec temple, an early Christian church, and the aforementioned Tlatelolco housing estate constructed under Regente Uruchurtu.

52. See Gutmann (2002) for a compelling account of how these events unfolded.

53. This number has been highly disputed since the event. While government officials were quick to publicize low figures (between thirty and forty, the president stated in the days following (Gutmann, 2002)), contemporary activists dispute even the figure of three hundred, claiming that the real number was much higher. I have heard the figure placed as high as two thousand dead in casual conversations, though such estimates are extreme and should be taken to illustrate the significance of the ideological role the memory of these events hold in current political imaginings (see Crane 2015), rather than a realistic assessment, though the obvious nature of the PRI government's attempts to cover up these events and dampen their effects make other figures above the consensus three hundred difficult to precisely dispute. See also Dresser (2014).

54. See Poniatowska's (2012) classic oral history collection *La Noche de Tlatelolco* for a compelling series of accounts differing markedly from official (and many academic) narratives surrounding the events and enduring legacy of the student movement and the massacre at Tlatelolco.

55. See Forero (2006); Dresser (2014); Martínez (2015).

56. The country at large followed a similar trajectory during this period, though the expected rate changes seem most starkly on display in the country's urbanized areas (see Tuiran et al. 2002; Haber et al. 2008).

57. Spending on education, for example, was more than cut in half between 1981

and 1987 (Haber et al. 2008). See also Escobar Latapí and González de la Rocha (1995); Centeno (1997).

58.   See Alarcón and McKinley (1998).

59.   See Álvarez Bejar (2016).

60.   A passive revolution entails, for Soederberg (2001, 105), "a change in forms of political and ideological domination by the ruling classes that is rooted in the wider restructuring of social relations of capitalist production, or civil society." PRI leadership followed a logic Soederberg refers to as "debt as discipline," whereby the state's far-reaching austerity measures and privatization programs would ultimately serve to draw from the Mexican economy the efficiencies thought to reside in the mysterious mechanisms of the market, and further used the arms of the extended state—such as educational institutions and media platforms, in addition to the sectors and organizations of the PRI itself—to propagate this logic and elicit the popular consent that forms the necessary complement to coercion. Soederberg's account of this process of reform and ideological reorganization, however, perhaps misrepresents the degree to which the neoliberal passive revolution took hold within the PRI itself. Davis (1994) argues that de la Madrid pushed for a broader set of democratic reforms both within the PRI (some of which would later take hold) and, more significantly, within the confines of the Federal District. Notably, Davis argues that de la Madrid intended to pass constitutional reforms that would have allowed for the direct election of the Federal District's Regente. That this provision was never made, even after several years of behind-the-scenes debate, provides evidence of the deepening rifts between party factions.

61.   Noting the difficulty of making robust longitudinal comparisons as a result of unreliable data, Haber et al. (2008, 197) nevertheless note "a broad consensus . . . that the distribution of income improved significantly between the early 1960s and the early 1980s and then deteriorated over the next two decades." The language of improvement can appear confusing here, especially given the paradoxically growing share of national income garnered by the upper-middle classes and simultaneous growth in income inequality overall (Hanson 1971) during the Miracle decades, as discussed above. Much of the yo-yoing growth and decline cited here is directly attributable to the middle classes (writ-large) of Mexico City, whose overall economic fortunes improved markedly during the Miracle and plummeted thereafter, especially during periods of marked austerity. As noted above, Ward (1990) demonstrates that though aggregate numbers therefore demonstrate a more equal share of income overall during the height of the ISI boom, some of the prosperity enjoyed by middle classes came at the expense of the lower classes (as well as those above). This arguably (and paradoxically) exacerbated the lived experience of inequality in parts of the city and further degraded the economic realities of the urban poor, in spite of the hopeful tone of aggregate figures. See also Alarcón and McKinley (1998); Angeles-Castro (2011).

62. As Roberts (2005, 114) argues, there has long been a "high degree of informal-
     ity of the Mexico City labor market," which, as in other Latin American cit-
     ies, decreased to some degree during the height of ISI policy, but expanded
     dramatically again in its wake. Rakowski (1994, 4) offers a usefully thorough
     definition of informality, which includes: "small-scale firms, workshops, and
     microenterprises with low capital inputs where production levels depend on
     intensive use of labor; nonprofessional self-employed, subcontracted put-out
     workers, disguised wage workers; unprotected or only partially protected work;
     illegal contractual arrangements, not fully regulated or registered or extralegal
     activities; activities that escape standard fiscal and accounting mechanisms;
     domestic service; cooperatives and associated activities with little or no sepa-
     ration between labor and ownership of the means of production; casual trade,
     street vendors, and market sellers, regardless of the source of goods; direct sub-
     sistence production." "Informalization," Rakowski continues, "refers to the cir-
     cumventing of regulations, benefits, payment of taxes, and so on by employers
     and the unequal and selective application of such by the state." The prevalence
     of such informality in Mexico City's economy is beyond dispute, though its
     precise share is practically impossible to estimate. But whether in the wealthi-
     est of hillside mansions or the most desperately poor of vecindades, informal
     transactions are so commonplace as to be entirely inconspicuous. This would
     include such service activities as cleaning and childcare, valet parking (often
     extra-legal, and often compulsory), and petty commodity production, and
     also the upscale businesses in gentrified neighborhoods that nevertheless must
     maintain budget lines for monthly or yearly bribes to local officials (in addition
     to those extra-legal expenses reserved for required licenses and permits), and
     quotidian business transactions that include casual reference to well-known
     avenues for avoiding legal tax requirements—sometimes to the extent of pre-
     paring several invoices—all of which is largely considered entirely mundane
     in my experience. I make use of the term here not in an effort to enter the de-
     bates on the functional utility of this sector vis-à-vis other categories of pro-
     ductive activity (see Rakowski [1994] for an informative survey of such infor-
     mality debates), but rather simply to highlight the difficulty of working with
     aggregate economic data under conditions where distinctions between formal
     and informal, legal and illegal, (etc.) are so difficult to parse in the course of
     daily economic and social life.
63. See Alarcón and McKinley (1998); Haber et al. (2008).
64. I discuss these events in the singular (i.e., the earthquake), following local
     convention.
65. Much of the city sits where a series of interconnected lakes had been until
     drainage projects (see Candiani 2014; Vitz 2018) over the past several cen-
     turies cleared the way for geologically precarious human settlement. At the
     edges of the former lakebed lie basalt rock surfaces, which also underlie much
     of the contemporary city (especially to the south and west). As a city official

explained to a group of my students in the Spring of 2019, the city therefore shakes as if it sat upon a great bowl of gelatin, and is susceptible to seismic events originating hundreds of kilometers away, as in the case of the 1985 earthquake (the epicenter of which was located on the country's Pacific coast).

66. See also Mier, Rocha, and Rabell Romero (1988).

67. Though some authors would claim in the ensuing years that criticism of the PRI state's response was overblown, and that it did make considerable effort in the aftermath, at least at the level of policy (see Gilbert 1998).

68. See Monsiváis (1987); Davis (1994); Poniatowska (1995); Gutmann (2002); Bruhn (2008); Ai Camp (2000, 2014).

69. Superbarrio dresses in the style of a lucha libre wrestler, and has appeared in this guise at many of the group's demonstrations. According to Gilbert (1998, 140), Superbarrio offered the following description of himself in 1991: "Of course, comparisons have been made with Superman. But he derives his power from supernatural, non-intellectual sources. My power comes from Mexican popular culture and the popular imagination. Superbarrio doesn't face vampires or wolf-men; he faces flesh-and-blood landlords, politicians and bureaucrats, and he has no kryptonite to protect him." His Facebook page can be seen at https://www.facebook.com/superbarrio.gomez

70. Such measures included things like rent regulations, density requirements, and various other favorable planning decisions and programs, such as the aforementioned pro-middle-class policies of Regente Uruchurtu (see Davis 1994, 2009), including the Tlatelolco housing projects.

71. There are important caveats to be made about such solidarities. In the course of fieldwork, I was told many times of the importance of the earthquake and the movements and groups that formed in its wake to grassroots politics in Mexico City and to the democratic transition (discussed later) in particular. I've sat and spoken with people who get emotional talking about their experiences of that time, or those of loved ones or even strangers. These emotions, though, are not organized for political convenience. There is sadness, grief, heartbreak, and even, perhaps, a measure of despondency that lingers after the losses suffered and traumas experienced. There is anger, even rage, over injustice and betrayal, which is, as much scholarship suggests, aimed at the state (and the PRI more especially). There is also affection, gratitude, and appreciation; families, neighbors, and communities found one another in the aftermath, and the movements and groups they formed would certainly propel radical political change in the years to come. But though community, solidarity, and resistance are extremely common themes across accounts, even these are far from straightforward. I have been told numerous times—often by people whose inflections, hesitations, and quasi-voluntary kinetic gymnastics I have largely seen as evidence of the difficulties of working through these memories and of telling clear stories about that time, especially to a stranger who wasn't there—of pain and frustration over defection, cooptation, corruption, graft,

and other real or perceived betrayals from among the *damnificados* themselves and those that claimed to speak on their behalf in the wake of the earthquake. Some of these wounds have healed through reconciliation (I was told) or simply through the passage of time, though many remain open or traumatically emerged afresh with the morbid anniversarial arrival of the 2017 earthquake. These experiences, too, refract contemporary sentiments about local leaders and political structures, and their influence can sometimes be found clinging to rumors and accusations of abuse, exploitation, ineptitude, duplicity, (etc.) aimed at local figures of the state and civil society alike.

72. See Azuela (2007); Bruhn (2008). The broader question of total or partial independence from the state and political parties continues to be a seriously debated issue among Mexico City's activists and civil society organizations. Many remain deeply distrustful of the state and of formal party structures, having witnessed the effects of cooptation strategies in particular that often result from cooperation with either. There was always, of course, a familiar third choice: cooperation with or cooptation by the PRI. As Azuela (2007, 161, original emphasis) writes of de la Madrid's expropriations in early October, "Overnight, a social movement that had brought some 150,000 people to the gates of the presidential house, adopted the shape of a disciplined queue in front of whatever state agency was to be created in order to *administer* the reconstruction."

73. According to the United Nations (2017a; 2017b).

74. Early returns favored the metropolitan area of the capital (and therefore Cárdenas, as this would be his strongest geographical base of support), fostering speculation about de la Madrid's motivation in halting vote tabulation and the alleged crash of the electoral computer system under the oversight of PRI Secretary of Governance Manuel Bartlett (discussed below). See Cantú (2019).

75. Various iterations of the phrases *se cayó el sistema* (the system went down, or the system crashed) and *la caído del sistema* (the system crash) commonly convey the infamy of this event with bitter jocularity (for instance, the Netflix series *Narcos: Mexico* named an episode dealing dramatically with the 1988 election and its irregularities "Se Cayó El Sistema"), despite decades of plodding insistence by Secretario Bartlett that this phrase was never used by officials at the time.

76. General agreement now exists among scholars and observers that "the PRI engaged in widely in fraudulent practices" in the 1988 election (Ai Camp 2014, 224). As Bruhn (1997, 140) puts it: "Nearly all observers agree that fraud marred the July 6 presidential election. Although it is impossible to prove that fraud changed the outcome, the evidence clearly shows that substantial electoral fraud benefitted Salinas and that Cárdenas was the chief victim." This evidence includes descriptions of armed robbery of ballot boxes, packets of ballets that did not match the numbers noted by poll watchers, partially burned ballots marked for Cárdenas found in rivers and alleyways, and voter registry irregularities and inaccuracies as high as 49 percent in some areas. Gutmann

(2002, 8) likewise characterizes this election as "what some have termed the most fraudulent presidential vote in Mexico's long history of corrupt electoral politics."

77. This task was specifically assigned to the Chamber of Deputies.

78. See Bruhn (1997); Ai Camp (2014).

79. Quoted in Smith (1989, 407).

80. See Inclán (2018).

CHAPTER 2

Epigraph. "It is the success of political reform that propelled the PRD; it is the first time in the country's history that there is a reliable electoral system, that there will be an autonomous government in the capital; it is the end of PRI hegemony, and for the first time, the system of checks and balances created in the Constitution of 1917 will begin to function."

1. As opposed to the PAN's relatively poorer showing with Clouthier in 1988, at just over 17 percent.

2. The lower chamber of Mexico's bicameral legislature, in which, after the 1988 elections and for the first time since its founding, the PRI did not hold the supermajority required to unilaterally alter the 1917 constitution. An unexpectedly strong showing in the 1994 elections, however, would allow the PRI to reclaim this privilege until the 2000 elections.

3. Alongside prolific privatization efforts (including the 1992 amendments to Article 27 of the constitution that made alienable property of many ejido lands across the country (see Siembieda 1996), undoing on yet another front the legacy of Lázaro Cárdenas) and other neoliberal reforms, Salinas announced a new national program to combat poverty at the outset of his sexenio. The Programa Nacional de Solidaridad (National Solidarity Program, known by the acronym PRONASOL, or more popularly as Solidaridad) targeted the poorest and most marginalized communities in the country. Widely regarded as a failure (see Moguel 2007; Enciso L. 2011; Fernández-Vega 2018), Zedillo reframed and expanded Solidaridad into the better known conditional cash transfer program Progresa, later rebranded again as Oportunidades under PAN President Vicente Fox (and maintained by his successor, Felipe de Jesús Cálderon Hinojosa) (see Soares et al. 2007; and see Campos, Esquivel, and Lustig 2012 on the trajectories of inequality in Mexico from 1989 through 2010) and again as Prospera under PRI President Enrique Peña Nieto in 2014.

4. Especially after further reforms in 1996, which increased IFE's autonomy.

5. Suspicion continues to swirl around this event, though Mario Aburto Martínez was tried and convicted of the crime, and claimed to be acting alone. Colosio's assassination came one day after former DF Regente Manuel Camacho Solís ended months of speculation surrounding an internal struggle for presidential candidacy by publicly announcing that he would not seek the party's nomination (Perez-Pena 1994). At the time of the Colosio's death, Camacho

was serving as chief negotiator to the Ejercito Zapatista de Liberación Nacional (Zapatista National Liberation Army, commonly referred to as the Zapatistas, hereafter EZLN), who had started an armed uprising in the State of Chiapas in the south of the country earlier in the year. Camacho had for years been the party's presumptive nominee (see Ward 1989). Speculation surrounding Camacho's possible involvement stemmed from eerie predictions of Colosio's political demise, as DePalma (1994a) reported, "Within a few weeks of the [EZLN] uprising, Mexico City newspaper columnists were writing plots in which Mr. Colosio could become mysteriously ill and would have to be replaced as a candidate by Mr. Camacho." The Mexican Constitution of 1917 bars any presidential candidate from holding political office for six months prior to standing for a presidential election, which cut severely into PRI leadership's choices in the wake of Colosio's murder, a fact which only further contributed to the rumors and suspicions of conspiracy. Zedillo, incidentally, had been Colosio's campaign manager prior to the assassination.

6.   In September of 1994, PRI Secretary General Francisco Ruiz Massieu was shot to death in his car by a machine-gun-wielding assassin near the Monument to the Revolution in the Federal District. Brother-in-law to President Carlos Salinas, Ruiz Massieu was reportedly being considered for minister of the interior, a powerful cabinet position, in then-President-Elect Zedillo's administration. The investigation into his murder became a major scandal, especially when Ruiz Massieu's brother Mario (also brother-in-law to President Salinas) resigned his post as assistant attorney general and chief investigator into his brother's assassination—having been appointed by President Salinas—alleging a cover up by the highest ranking officials of the PRI, and fled the country. He was intercepted on his way to Spain in Newark, New Jersey and arrested for carrying US$46,000 in undeclared cash. Facing money laundering charges in Texas, Mario Ruiz Masseieu died of an overdose (and apparent suicide) while under house arrest in New Jersey in September, 1999. He had reportedly made twenty-six deposits—largely via proxy—of between US$40,000 and $800,000 each in a Houston, Texas bank over a fourteen-month period from 1993 to 1995, totaling US$9,900,000, while serving as assistant attorney general, a post in which he was charged with prosecuting major narcotics traffickers (Gunson 1999). President Salinas's own brother, "businessman" Raul Salinas de Gortari—who the US General Accounting Office reported to the US Senate Permanent Subcommittee on Investigations (within the Committee on Governmental Affairs) had illegally transferred some US$90,000,000–US$100,000,000 to personal Swiss bank accounts between 1992 and 1994 with the aid of Citibank (a subsidiary of Citigroup, on whose Board of Directors Zedillo currently sits, as of August, 2020) in New York City (USGOA 1998)—was found guilty of masterminding the Ruiz Massieu assassination and sentenced to fifty years in prison (he was released when his conviction was overturned some ten years later). Another alleged conspirator, Congressman

Manuel Muñoz Rocha, disappeared in the days following Ruiz Massieu's assassination, and it was suspected that it was his remains which were found two years later on a ranch outside Mexico City owned by Raul Salinas, who claimed that federal prosecutors had planted them there (Fineman 1996). See Castillo Garcia (2001). See also DePalma (1994b) for a brief summary of several other scandals and otherwise unhappy events of 1994 in Mexico City.

7.   In February of 1998, the *Washington Times* cited a leaked CIA document in accusing Labastida of dubious connection to narco-trafficking in Sinaloa, where he had served as governor from 1987 through 1992. Zedillo and others publicly defended Labastida—who was far from the only prominent Mexican politician to be accused by the US press of unsavory narco dealings during the 1990s, see Chabat (2002); Esquivel (2020)—and the accusations and unwelcome attention were not enough to prevent Labastida's nomination. See also Cason and Brooks (1998).

8.   Part of this pivot toward political territory, Langston (2017) explains, was due to the decline of union membership (and thus organized labor's ability to reliably deliver votes for the PRI) because of economic restructuring associated with neoliberal reforms. See also Foweraker and Landman (1995).

9.   See Fernández Santillán et al. (2001); Franco, Schteingart, and Ugalde (2015).

10. See Lomnitz (2003) on this usage and its meanings.

11. These largely took effect in the fall of 1997, aside from the direct election of the delegados (executive heads of the Federal District's sixteen delegaciones), which did not take effect until 2000 (Becerra Chávez 2003).

12. Cárdenas would vacate the office in late September of 1999 in order to run for the national presidency again in 2000, leaving the office to then–PRD Secretary General María del Rosario Robles Berlanga, who was reportedly labeled the ideal candidate by Cárdenas's inner circle (Olayo and Llanos 1999). Robles, whose corruption scandals would cause major damage to the PRD's reputation during her brief tenure as mayor and subsequently in various federal posts, was arrested in August of 2019 and remains (as of May, 2021) incarcerated awaiting judgement on misconduct and the mishandling of public funds to the tune of some 5–7 billion pesos in a scheme known as la Estafa Maestra. Early in March of 2021, Robles offered to proffer a confession in exchange for sentencing considerations, though this offer was retracted some six days later, citing a lack of faith in the judicial process and accusations of unjust persecution, including their belief in authorities' intention to "keep her in prison at any cost and under any circumstances" (quoted in Varela 2021). See also Castillo García (2020); Vela (2021).

13. For a general discussion, see *The Economist* (2012). See also the pioneering and controversial work of Thompson (1929).

14. Damnificados is the name given to those displaced by the 1985 earthquake. See Gerlofs (2021) for a brief discussion of differing figures and their significance.

15. I was repeatedly offered this impression of Centro in particular in both formal

interviews and casual conversations. The general portrait most typically offered is of drastic decline after the earthquake, when those with the ability fled the destruction and the period of physical and political tumult that followed. This legacy has been, only somewhat unfairly, remarkably resilient. In the first few months of my fieldwork in 2015, I was repeatedly and fervently advised that Centro remains a no-go zone after dark.

16. The bloating disutility and creeping Anglocentrism of this language have received considerable attention in recent years. See Ghertner (2015); Smart and Smart (2017). See also Wyly (2015); López-Morales, Bang Shin, and Lees (2016); Slater (2021).

17. Ackerman (2014) argues that this "PRIAN" alliance—tantamount to a coalition government—was never seriously challenged (despite a moment of PAN-PRI cooperation in 1996–1997), relegating claims of a democratic transition to the status of myth.

18. See Tuckman (2012a).

19. See Ackerman (2012).

20. One such interested party was billionaire telecom and real estate magnate Carlos Slim, who began amassing property in Centro and the area that would come to be known as Nuevo Polanco (also sometimes colloquially called Ciudad Slim) in the early 2000s. Slim's contributions to the transformation of Mexico City are briefly discussed in Chapter 4.

21. See Grayson (2007); Becker and Müller (2013); Leal Martínez (2020).

22. In local parlance, this decidedly pejorative phrase typically connotes a group of celebrity architects, property developers, financiers, landlords, and public officials acting as a sort of cabal in urban redevelopment projects.

23. In my experience, contemporary expressions of such grievances are unevenly—and perhaps rather unfairly—distributed among PRD mayors. AMLO's extremely controversial decisions and relationships, for instance, receive far more ambiguous (and cautious) attention than those of Ebrard or Mancera. During my extended fieldwork (2015–2016), specific allegations against AMLO on this front (e.g., coziness with developers, displacement-inducing initiatives, misplaced priorities) were sometimes taken seriously, and sometimes met with deflection or outright—and rather obstinate—dismissal, often through narratives of political persecution and/or unreasonable scrutiny inflected in various ways by persistently familiar notions of conspiracy (see, for instance, Rosales 2018; Hiriart 2019; *La Jornada* 2021).

24. See Crossa (2009); Kanai and Ortega-Alcázar (2009).

25. Most notably the events surrounding the expropriation and development of a valuable piece of land in the Santa Fe area known as El Encino, and the subsequent associated *desafuero* saga of 2004–2005. Less than a month before AMLO took office in December of 2000, Rosario Robles's administration expropriated a parcel of land near Santa Fe—by then already a ritzy edge city on the southwestern side of the metropolitan region—in order to construct an

access road for the private ABC Hospital. Construction was ordered halted by Federal Judge Alvaro Tovilla León less than a month later, who later found that construction nevertheless continued on the site in the ensuing months. In response, Fox and his allies (importantly, a coalition of both PAN and PRI legislators) voted to strip AMLO of the immunity from prosecution guaranteed under the 1917 Constitution (in a process known as a desafuero), on the grounds of abuse of executive authority. Largely considered an effort to keep AMLO out of the 2006 presidential race (as Mexican citizens were at the time stripped of their civil rights while under federal prosecution, candidates were barred from entry in federal elections—one such right—if facing federal charges), the highly publicized desafuero process instead ultimately rallied loyal supporters and garnered international attention for AMLO's campaign. A series of editorials in major domestic and international news outlets decried what many considered a naked attempt to keep AMLO from the presidential race, and demonstrations mounted in the capital and elsewhere in support of the embattled mayor, especially after the announcement of the desafuero's passage in April of 2005 (see Salgado and Bolaños Sanchez 2005; Lopez and Aviles 2005; Proceso 2005). After briefly stepping down to await arrest on April 8, AMLO resumed the office on April 25 in the wake of mass demonstrations and assurances that he would not be arrested by the Procuraduría General de la República (attorney general, hereafter PGR). In a remarkable show of defeat, Fox removed General Rafael Macedo de la Concha (also a military general) as attorney general just two days later on April 27, and seven days later announced (through new appointee Daniel Cabeza de la Vaca) that no charges, after all, would be pursued against AMLO. See Grayson (2007) for a clear timeline and thorough accounting of these events.

26. Typically referred to in English as the Federal Electoral Tribunal. The TEPJF was created by the aforementioned reforms of 1996, superseding an earlier tribunal created by the 1990 electoral reforms, alongside the IFE. This body performs the function of certifying federal elections previously performed by the Chamber of Deputies (as in the 1988 election, discussed in Chapter 1).

27. After a limited recount of some 9 percent of precincts, final results declared AMLO's official total short of Calderón's by 233,831 votes, or 0.56 percent.

28. Becerril et al. 2006.

29. "Votes were merely symbolic," Grayson (2007, 268) notes, "because the Zócalo holds only 110,000 people, a fraction of the number of convention delegates."

30. See AMLO's November 20 remarks, printed the following day in *La Jornada* (López Obrador 2006). Notable among the members of AMLO's cabinet is the appointment of Claudia Sheinbaum Pardo as secretaría de la defensa del patrimonio nacional (secretary of national heritage). Sheinbaum was a co-recipient of a Nobel Peace Prize for work on climate change (as part of the Intergovernmental Panel on Climate Change) in 2007, and was elected jefe de gobierno of Mexico City in 2018.

31. The PRD also found its share of the legislature on the decline in 2009, after positive previous cycles. According to INE, PRD seats in the five-hundred-member Cámara de Diputados increased in both 2003 and 2006, at the expense of both the PRI and the PAN. From a deep nadir of 50 seats in 2000 (against 211 for the PRI and 206 for the PAN), the PRD held 96 seats after the election of 2003 (against 225 for the PRI and 152 for the PAN) and 126 after 2006 (versus 104 for the PRI and 206 for the PAN). The party's share then dipped again to just 71 after a poor showing in 2009, while a resurgent PRI sat 237 (and the PAN 143). The pattern is less pronounced in the 128-seat senatorial chamber (with sexennial elections), with the PRD's share increasing from 16 in 2000 (against 60 for the PRI and 46 for the PAN) to 29 in 2006 (against 33 for the PRI and 52 for the PAN), then dipping to 22 in 2012 (against 54 for the PRI and 38 for the PAN). These numbers are complicated by the cyclical entrance and exit of newly represented parties benefitting from the electoral reforms of the 1990s, such as the Partido del Trabajo (typically called the Labor Party in English, hereafter PT, formed in 1990).

32. See Crossa (2013); Müller (2013); Davis (2014); Rasmussen (2017). *Artisan* and related designations are highly contested language among vendors, craftspersons, and other ambulantes, as explored by Crossa (2016).

33. Though with important caveats, discussed in Chapter 4.

34. The first elections for the ALDF were held in 1997. The PRD won 44.19 percent of the votes (against 23.06 percent for the PRI and 17.66 percent for the PAN) in 1997 elections, 30.73 percent in 2000 (against 22.00 percent for the PRI and 35.03 percent for the PAN), 43.29 percent in 2003 (against 11.51 percent for the PRI and 25.03 percent for the PAN), 49.92 percent (as the headliner of the Por el Bien de Todos coalition with the PT and Convergencia, calculated aggregatively by the IEDF in 2006) in 2006 (against 12.76 percent for the PRI (as the headliner of the coalition Unidos por la Ciudad) and 24.98 percent for the PAN), 21.66 percent in 2009 (against 15.99 percent for the PRI and 19.74 percent for the PAN), and 32.60 percent in 2012 (against 16.66 percent for the PRI and 18.51 percent for the PAN). The 2009 election is notable for a major increase in null votes (10.51 percent), which had never before topped 3 percent, according to the IEDF.

35. During and after the elections, AMLO and others accused the PRI of widespread electoral fraud, including vote buying. Of these accusations, arguably the most relevant in Mexico City was that publicly levied against major grocery and department store chain Soriana, the second largest grocery retailer in all of Mexico behind only Walmart (and its subsidiaries). Voters were allegedly offered Soriana gift cards in exchange for PRI votes, a charge both the PRI and Soriana have always denied. Evidence of this practice abounds in testimonial and video forms, as do personal stories, rumors, and jokes of Soriana-aided vote buying gone wrong, as when voters, often without any sense of irony, accused operatives of distributing faulty or empty gift cards they were later

unable to use to buy groceries. In January of 2014, IFE declared the accusations of electoral fraud unfounded (*El Economista* 2014). The allegations nevertheless loom large in the city, and I have met several anti-priistas in Mexico City who remain committed to a (likely inconsequential) boycott of Soriana stores as a result. See also Tuckman (2012b) for these and other allegations of voter fraud in the 2012 election, which, she claimed, "exposed these practices as never before thanks to the explosion of mobile phone ownership and social media, as well as the activism of the new anti-PRI student movement that erupted in May [2012]."

36. As a movement, Morena had been part of the PRD coalition supporting AMLO in the 2012 elections, but was converted into an official political party in the months following. The name Morena draws on the feminine form of *moreno/a*, which is primarily used to refer to a dark-skinned (Indigenous) person, and is intentionally suggestive of the party's political orientation. As the name ends in *a* rather than the masculine *o* (as in *moreno*), it connotes a feminine person. The theoretical referent would be arguably be the most marginalized person in Mexican society: a dark-skinned woman.

37. Tragically, Tuckman died in Mexico City at the age of fifty-three in July of 2020, having been diagnosed with cancer the previous year.

38. As Davis (1994, 320) thoroughly demonstrates, "far from the monolithic, Machiavellian party with absolute and willful control over policy and civil society, the PRI has shown itself to be a heterogeneous concoction of social classes and political factions holding little consensus over critical issues, especially urban services and administration in the capital." See also Joseph and Buchenau (2013).

39. A danger best exemplified the 2017 earthquake, which struck on the same date as that in 1985, causing severe damage to many of the same neighborhoods.

40. Following Williams, who famously (1977, 87) conceptualized a relationship of determination as a "setting of limits" and an "exertion of pressures."

CHAPTER 3

1. Though this history is somewhat murky, some claim that some MUP leaders used their influence in the post-earthquake years to enrich themselves and/or develop their own patronage networks, channeling and directing government assistance according to their own agendas. Such actions are understandably said to have created conflict among leaders of different factions of the MUP.

2. Though with at least one important caveat, to be discussed below.

3. There is some confusion on this point evident in the international discussion of the Charter. Adler (2015a; 2015b), for instance, incorrectly ascribes the force of law to the Charter and describes its character as that of legislation. Its actual status as a legal entity will be discussed in detail below, but it should be noted that it holds no formal place in law, despite the public endorsement of Mayor Marcelo Ebrard and other members of the GDF in the summer of 2010 and subsequently.

4. Among many others, see Brown and Kristiansen (2009); Fernandes (2007); Gray and Porter (2018).

5. See Holston (2009) and Vradis (2012) for examples of the unexpected use of rights talk by organized criminal gangs in Brazil and a resurgent political right orienting itself toward the city in Greece, respectively.

6. My usage of *dialectics* and *the dialectic* follows Ollman (2003). Formative guides to dialectics are also provided by Ollman (1971, 1993), Jameson (2009), and, to a lesser extent, Walker (1977).

7. An exchange between protagonist Alice and the Queen in which this queer sort of memory is elaborated by the latter is said to have been a favorite episode of Carl Jung, progenitor of synchronicity in the field of psychoanalysis.

8. See Harvey (2006).

9. See also Harvey (1974); Henderson (2013).

10. "A spider constructs operations that resemble those of a weaver, and a bee puts to shame many an architect in the construction of her cells. But what distinguishes the worst architect from the best of bees is this, that the architect raises his structure in imagination before he erects it in reality. At the end of every labour-process, we get a result that already existed in the imagination of the labourer at its commencement" (Marx 1967a, 174).

11. In "Becoming and the Historical," for example, Lefebvre (2003b, 66) explains his methodological quest for becoming thus: "ternary or triadic analysis grasps becoming (or at least comes nearer to it than the rest). . . . The Marxist triad 'statement-negation-negation of the negation' aims to produce becoming, but so far that high ambition has not been fulfilled, in particular as far as the State and the community (which was negated by history and re-established under communism) are concerned."

12. This triadic exploration of mediation, centrality, and difference in Lefebvre's right to the city follows a similar treatment by Schmid (2012).

13. For instance, through the realization of military and police operations or financial geopolitics conceived in urban "centres of decision-making" (Lefebvre 1996) or the production of tradable futures and the rapid centrifugal spiraling of related credit instruments exponentially distanced from actual commodities, respectively. See also Moreno (2014).

14. See Parnell and Pieterse (2010); Fernandes (2007). See also Purcell (2002) for a nearly ad nauseum hypothetical extension of the right to the city as the right to participation in urban governance.

15. See Williams (2002); Merrifield (2006).

16. See Merrifield (2006, 113), who argues that for Lefebvre, "the right to difference cried out as loud as the right to the city."

17. The categories of "the deprived" and "the discontented" include respectively: the "excluded," the "working class," and the "directly oppressed"; and the "small business people," the "gentry," the "capitalists," the "establishment intelligentsia," the "politically powerful," the "alienated," the "insecure," the "hapless lackeys of power," and the "underwriters and beneficiaries of the established

cultural and ideological hegemonic attitudes and beliefs" (Marcuse 2009: 190–91).

18. See Park and Burgess (1925). See also Ford (1996) for a treatment of Latin American cities in similar fashion.

19. "Between equal rights, force decides," Marx (1967a, 225) stated in his discussion on the working day. Mitchell (2003) argues (following Harvey) that this does not constitute a wholesale rejection of the language or institution of rights, per se, but merely an insistence on accompanying force.

20. For instance, Hearne (2014, 15) insists that "debate centres on defining the right to the city," and that "[a] central task for urban scholars and practitioners is to work out not only what the right to the city means, but also how it can be practically achieved." Purcell (2014, 142) argues that "multiple, specific formulations can engender a sustained debate about the best way to understand the idea." In contrast, I argue for a plural understanding of the right to the city (see also Pierce Williams, and Martin 2016) not to find or forge a "best" version as theorists or practitioners champion competing visions, but rather both because this is how the idea functions best methodologically and because of the strategic possibilities such a conceptualization reveals in practice, as elaborated below.

21. As illustrated in the above epigraph from Borja et al. (2006).

22. See Brown (2013) for a brief history of the right to the city in the context of these meetings, and the role played therein by HIC-AL.

23. In October of 2015, Ortiz was awarded the National Prize for Architecture in a grand ceremony at the Palacio de Minería in Centro. Having been privileged to attend at the invitation of HIC, I witnessed Ortiz's humble acceptance of this great honor, choking back tears as his life's work was celebrated by his peers and collaborators, by political figures who had been allies and adversaries (including Mayor Miguel Ángel Mancera), and by the city's most vulnerable residents whose conditions it had been his longtime ambition and cause to improve. Some of these last had brought handmade signs to the otherwise debonair affair, such as those depicted in Figure 3.1 which together read, "Enrique, thanks to your social struggle, I have a home" (Enrique, gracias a su lucha social, tengo un hogar).

24. It is also worth noting that the Spanish word *derecho* has several meanings in Mexico (as elsewhere). It can mean "right" or "duty" in the juridical sense; "right" in a moral, ethical, or logical sense (e.g., "the right thing to do," "upright," or "honest" when applied to a person, group, or institution); "law" (writ-large); "straight" in a locational sense (e.g., "straight ahead"); or "straight" in a mixological sense (e.g., a "straight" mezcal, which only a fool would order otherwise).

25. Such as Casa y Ciudad and the Centro Operacional de Vivienda y Poblamiento (Operational Center for Housing and Population, hereafter COPEVI), among others.

26. Adler (2015a) commits an analytical misstep in quoting Rello thus: "The

Mexico City Charter for the Right to the City, without a doubt, is the clearest instrument to continue our long-awaited dream . . . a city of rights for everyone. There is no turning back." Aside from some important differences of translation, this omission of a crucial portion of Rello's sentiment is exemplary of a common problem in interpreting these events, that of glossing over or ignoring entirely the role of the impending political reform in the history of the right to the city in Mexico City. To miss or ignore this crucial factor is to leave little beyond coincidence to explain why the Charter was finally developed after some twenty years of advocacy, and why it gained such widespread and influential institutional and municipal support.

27. In 2008, the World Social Forum was decentralized, and was held simultaneously in multiple cities across the globe, including Mexico City. This was a departure from previous years in which a single city (or, in the sole case of 2006, three cities) played host to the event.

28. Anonymous, interview with the author, 2014.

29. As of September 2011 the Charter boasted 257 organizational signatories (aside from individual signatories, which are not listed in the printed version), grouped together by the CPCCMDC (2011) as: CPCCMDC members (5); academic institutions (6); guild societies (3); indigenous and peasant organizations of indigenous villages and communities (38); ejidos and villages (21); cooperatives (11); small commerce organizations and public workers (13); unions (7); transit associations (11); civil organizations (72); women's organizations (8); LGBTTI rights organizations (9); social organizations (41); and habitat unions (12). Interestingly, section 1.3 ("Territorial Ambit") addresses the Charter to the "Delegations of the D.F. and their urban and rural areas," and speaks of the need to increase powers of metropolitan coordination as part of the political reform process (CPCCMDC 2011). Indeed, it is telling that the right to the city was never framed as the right to city in the Federal District, but rather always as the right to the city in Mexico City, a legal entity which did not then exist. This is yet another illustration of the deep entanglement of the Charter's politics with those of political reform.

30. It is certainly tempting to call this organization of forces, institutions, groups, and persons a movement, especially in the wake of the Charter's endorsement and the dissolution of the CPCCMDC. From an academic perspective, this grouping would easily fit most definitions of an urban social movement, from the narrow, as in Castells's (1983) usage—wherein "urban social movement" status is reserved for mobilizations that are successful in garnering some measure of structural change relating to collective consumption—to the general, such as that of Nilsen (2009, 114), who defines a social movement as "the organisation of multiple forms of materially grounded and locally generated skilled activity around a rationality expressed and organised by (would-be) hegemonic actors, and against the hegemonic projects articulated by other such

actors to change or maintain a dominant structure of entrenched needs and capacities and the social formation in which it inheres, in part or in whole." Given its varied focus on issues not only of collective consumption, identity and community privileges, and other goals aimed directly at the state, the coalition falls uncomfortably among discussions of so-called new social movements; see Habermas (1981); Young (1990); or see critical reviews of new social movement theory by Beuchler (1995); Pichardo (1997); and the demand-less political movements Swyngedouw (2012) identifies. Whether or not this coalition is reasonably considered a movement by academic standards, however, is in many ways the wrong question. In Mexico City it is simply not described in this way. In this context, the Spanish word movimiento (movement) invariably refers to the MUP, which is considered *the* movement. Other descriptors are sometimes used, especially plataforma (which can be translated as "movement," but is in this case better translated as "platform," which can refer to a position, a document, or a group pursuing something such as a position or document), though I found it necessary in interviews and casual conversations to be quite specific when asking questions about the right to the city. Plataforma, for instance, could mean the coalition that gave rise to the Charter, the Charter itself, the group that still meets on a semi-regular basis to discuss further action in the wake of the Charter, or more likely it could refer to efforts to improve the Global Charter for the Right to the City, in which many of these same local persons are involved. Because the aim of this chapter is to understand the right to the city in its geographical and temporal origins and the wealth of its political and other entanglements rather than to make a categorical assessment of its advocates, I have opted to use the term *coalition* to describe their always fluctuating coherence as a set of social and political forces and personalities. Even this word presents some challenges, as the Spanish *coalición* can be used, and in some cases is, to refer to HIC-AL (Habitat International Coalition), though this organization is usually referred to simply as HIC (often laboriously pronounced "heek").

31. This initiative was pursued in large part by COPEVI, an organization also once headed by Enrique Ortiz Flores.

32. Brugada cited the public cost at 114 million pesos, a budget which was compared favorably to those of other city projects, such as Parque Bicentenario, a botanical garden inaugurated by the city in the distant northern borough of Azcopotzalco in 2010 and constructed at a public cost claimed to be in the billions (*GreenTV Noticias* 2012).

33. Accusations were later made against Brugada and her associates regarding the origins and possible mistreatment of the animals of the Cuitláhuac Zoo (Méndez 2016), and some animals were removed due to "irregularities" in their treatment and legal status (Excélsior 2016).

34. Though the delegado to follow Brugada in Iztapalapa, Jesús Valencia Guzmán,

reportedly had quite different priorities and commitments from those of his predecessor, he nevertheless continued to reference the right to the city in his praise of Parque Cuitláhuac.

35. Ebrard would call it "one of the most important works of engineering in the country and the world" during a tour of the highway in 2012, and claim that its soaring surface boasted a view "without equal in all the city" (Gómez Flores 2012).

36. The Supervía Poniente is often referred to as a private highway, which in the United States would be called a tollway—a roadway with limited and monitored access for which a toll must be paid—in this case based on how far a car travels. As of May 2017, the toll for the four-exit length of road was $MX63.

37. Mexico City daily *La Jornada* ran a story on the press conference announcing these results, to which a self-described lone wolf who supported the project and participated in the survey was the only resident attendee. When asked how many more local residents his presence should be understood to represent, he offered the following remarks, which *La Jornada* seems to suggest may also offer some insight into the survey's level of representativeness: "The quantity does not matter for me. We look for quality, more than quantity. Quality in what sense? Not in the economic sense, but that they are thoughtful people, who have an open mind, that are ready for change" (Salgado 2010).

38. Anonymous, interview with the author, 2015.

39. Anonymous, interview with the author, 2015. See also Quintero Morales (2017).

40. I witnessed several such accusations in public and semi-public forums. It came up more than once at the second CONDUSE meeting in December of 2015, when several people remarked on the mayor's conspicuous absence in such venues, and notably again in a meeting between the office of the city's secretary of government, in the person and staff of Undersecretary Juan Jose García Ochoa, and organized residents represented by members of the MUP. In this latter meeting, the undersecretary was aggressively asked what civil society groups would have to do to get Mancera to meet with them, a question that echoes a common refrain about the difference between the Ebrard and Mancera regimes. Political commentary and cartoons also frequently remark on and poke fun at the mayor's distance from his public, which stands in stark contrast to the image he sought to create in his mayoral campaign (an issue more fully addressed in Chapter 4). This is also seen as something of a departure from the image of the PRD more broadly, though even this association with the nominally leftist party has begun to fade as a result both of the rupture with AMLO and the perceived rightward drift of recent PRD leadership.

41. Anonymous, interview with the author, 2017.

42. Anonymous, interview with the author, 2015.

43. Marcelo Ebrard Casaubón, "Palabras de Marcelo Ebrard en la Firma de la Carta por el Derecho a la Ciudad." Noeve Sietesoles, YouTube, July 13, 2010, 4:14. https://www.youtube.com/watch?v=leBUSbWtPrw.

CHAPTER 4

1. See Bruhn (2008); Kohout (2009); Tuckman (2012a).
2. This cloak of democratic legitimacy is the most compelling reason for taking seriously Peruvian poet and Nobel Laureate Mario Vargas Llosa's famed 1990 characterization of PRI–Mexico as "the perfect dictatorship." Vargas Llosa would later praise Mexican democracy, mirroring the optimistic language of the *democratic transition*, as in the epigraph to this chapter.
3. See, for example, the *New York Times* (2000). See also Tuckman (2012).
4. See Dávalos (2004), for example. See also Nelson (2003); Gugelberger (2005); Lombard (2013).
5. *Sol Azteca*, or "Aztec Sun," is the party's nickname and main graphic symbol.
6. The law has seen several rounds of revision, the most sweeping coming in 1998 and 2010.
7. Plebiscites result in binding decisions of this sort, for instance. Results of other processes, such as referenda and consultations, do not. In the case of the referendum, the required action is stated thus (translation by author): "The results of the referendum shall not be of binding character for the Legislative Assembly, its effects shall only serve as elements of evaluation for the convening authority. The results of the referendum shall be published in the Official Gazette of the Federal District and at least one of the daily newspapers of major circulation" (IEDF 2013).
8. See Davis (1994); Gilbert and Ward (1984); Jimenez (1988). The field of planning is no exception, as Azuela (2007) demonstrates.
9. Ellingwood (2012) found these sentiments similarly pervasive in the city in the run-up to the 2012 elections, especially among "democracy babies," persons who grew up or came of age in the post-transition era.
10. See, for example, Johns (1997).
11. See Allmendinger and Haughton (2012); Swyngedouw (2009).
12. See Crouch (2004); MacLeod (2011); Rancière (1994); Swyngedouw (2011).
13. See Rancière (1994, 2010).
14. See, for example, Žižek (2008).
15. See Mitchell, Staeheli, and Attoh (2015).
16. See Derickson (2017); Dikeç (2017).
17. See Davidson and Iveson (2015); Swyngedouw (2009, 2011).
18. See Dikeç and Swyngedouw (2017).
19. See MacLeod (2011); Mitchell, Staeheli, and Attoh (2015).
20. See Derickson (2017); Dikeç (2017).
21. Levy should be understood to be speaking here of the planned project soon to be unveiled to the public, rather than about the avenue itself.
22. The boulevard was originally christened Paseo de la Emperatriz (Promenade of the Empress) in honor of Maximiliano's wife Carlota, daughter of Belgian King Ferdinand II. It was renamed by Juárez after his government returned

from exile in Veracruz after the collapse of the Second Empire and the execution of Maximiliano in 1867. The Castillo, shelled by the US army during the late stages of the Mexican American War, is now a popular tourist attraction offering excellent vantages of the city skyline and a window onto high society during various Mexican epochs.

23. The monument, crowned by a golden figure of Nike, Greek goddess of victory, sits atop a column emanating from the center of a structure containing a mausoleum which houses the remains of several heroes of Mexican Independence. On the Diana, see Leñero (2004).

24. Parnreiter (2015, 27) describes the rapid development of the Reforma area through the López Obrador and Ebrard mayoralties' use of planning instruments known as Corridors of Investment and Development and Corridors of Integration and Development (both abbreviated CID), by which the mayor's office was able to centrally control the planning of specific development sites, "flexibly created spatial entities which are defined by their economic potential."

25. I was told by numerous acquaintances that foreign corporations routinely rent blocks of apartments for their upper-tier foreign labor forces. I visited at least four such apartments during fieldwork.

26. *Tianguis* are local markets, usually of an informal character though often occurring on the same site at regularly scheduled intervals or, as in the case of the site described above, existing on a single site uninterrupted in a precarious semi-permanence. The word is of Nahuatl origin. Local residents anecdotally lamented to me that the market referenced here had been there for many decades, and that at least some of the vendors had been bribed to move their livelihoods elsewhere without making a fuss. One friend reported that, among other things, this market had been his favorite place to purchase bootlegged DVDs.

27. See, for example, Fuentes (2020).

28. This group could be said to include Simón Levy and ProCDMX, the Jefetura, several prominent architects, and the so-called real estate cartels, along with the retail groups whose stores routinely pervade such developments in a familiar pattern across the city.

29. Anonymous, interview with the author, 2015.

30. The language of the LPC states that should the results of the consulta differ from the plans of the concerned authority, the latter "shall be required to express with clarity the motivation and foundation of their decisions" in relation to the views expressed by citizen participants (IEDF 2013).

31. Saying that Romero was enjoying their moment in the sun, in 2011 Forbes labeled them "the world's richest man's favorite architect" (Dolan 2011).

32. Unknowingly illustrating the concerns of those who objected to the globalizing aesthetic of many of Mexico City's rapidly developing neighborhoods, The *Wall Street Journal* described approaching Museo Soumaya thus: "There's not a taco stand to be seen. Even if you've made the trip before, the slick

establishments paving the way are just enough to make you forget you're in Mexico" (Casey 2011).

33. The Metrobus is a recent addition (the first lines were introduced in 2011) to Mexico City's transit network. They run in generally straight lines across several sections of the city, and, unlike the countless smaller buses casting complex webs across the metropolis, run in lanes specifically reserved and barricaded off for their exclusive use. Metrobuses stop at dedicated elevated stations that sit in the medians of the city's largest streets, such as the Avenida de los Insurgentes.

34. This has long been a popular gathering spot for youths and members of Mexico City's counter-culture, along with serving as a common meeting place at the junction of several Metro and Metrobus lines. It also sits at the corner of several distinct neighborhoods. Its commercial spaces continued for years to be dominated by services increasingly rare in the city, such as those offering incremental internet access, though at my last visit to this space in July of 2019, all of these businesses were suddenly and quite conspicuously gone.

35. See especially *Código* (2015); Ruiz (2015).

36. See Medina Ramírez (2015); Villavicencio (2015).

37. "Resident" is my preferred translation of *vecino*, the term these people usually use for themselves. It can also be translated more simply as "neighbor," and less simply as "inhabitant." I have also heard it used and explained as something approaching "citizen," with an explicitly political dimension.

38. This latter group of *vecinos* display racialized and otherwise stereotypical attire and other aesthetic cues that mark them as decidedly "other," reflecting local residential trends (and anxieties). When the "true" residents object to Mancera and Levy, they are told that these people are indeed residents, as they have been granted voter ID cards.

39. Incidentally, these archways provided my own first exposure to the project, as I wandered through and around them alongside other park-goers in early October, some of whom confessed confusion about the project nearly as profound as my own.

40. The right to the city hardly figured in these efforts—indeed was glaringly absent—despite the fact that many of those active in the counter-campaign had been involved or at the very least familiar with the CPCCMDC and the Charter. The public contest over the Corredor project can thus be understood as a prominent example of the right to the city not traveling or being pursued under precisely this name, in spite of seemingly obvious resonance.

41. See Llanos Samaniego et al. (2015); Regeneración (2015); Torres (2015).

42. Providing a commentary on the establishment and cohesion of Cuauhtémoc, Blanco (2004, 222) writes, "The Cuauhtémoc District, like the rest of Mexico City's district boundaries, was a brain wave of President Luis Écheverria that our city hardly deserved. It was drawn with an arbitrary stroke of a highlighter over a city map, with no regard for history, economics, or demographics. Why

separate the Zócalo from the nearby La Merced market, yet lump it together with the distant Zona Rosa? . . . I don't see the Cuauhtémoc District as a harmonious entity, but rather as a ragbag of clashing locations."

43. Maps of the vote seem, at least partially, to substantiate this suspicion, as these neighborhoods, which are overwhelmingly low-income, showed a pattern of voting for the initiative at a much higher rate (as high as 71 percent in favor) than their neighbors (most of which voted at least 50 percent against). These neighborhoods betray no other obvious reason this trend, and no competing explanations of this trend have achieved even reasonably wide circulation in the city. One such map of the vote and rates of participation can be found at: http://subversiones.org/archivos/120397.

44. Anonymous, interview with the author, 2016.

45. Anonymous, interview with the author, 2016.

46. It should be pointed out, too, that many residents I spoke with voiced little concern with Mancera's reasoning, and were happy to simply enjoy the favorable result.

47. As explored in Chapter 3. Indeed, both Ebrard and AMLO presided over the construction of such second-level highway projects in blatant and highly publicized contravention of citizen outrage, a history not lost on opponents of the Corredor project and often mentioned in their media campaign.

48. According to the IEDF, Mancera received 63.58 percent of the vote, Paredes received 19.73 percent, and PAN candidate Isabel Miranda de Wallace received a meager 13.61 percent.

49. In 1997, Cárdenas received 48.1 percent of the vote, while the closest competitor, Alfredo del Mazo (PRI) received 25.6 percent, and Carlos Castillo Peraza (PAN) came in third place with 15.6 percent (Emmerich 2005). For further comparison, AMLO was elected by a margin of only 5.57 percent (AMLO received 39.5 percent and Santiago Creel (PAN/Verde) and Jesús Silva Herzog Flores (PRI) received 34 percent and 22.7 percent, respectively, according to *La Jornada* (2000)), and Ebrard by a margin of 19.11 percent (Ebrard received 46.37 percent, with Demetrio Sodi (PAN) and Beatriz Paredes Rangel (PRI/PVEM) receiving 27.26 percent and 21.59 percent, respectively, (IEDF 2006).

50. See Žižek (2008).

51. The postpolitical ultimatum par excellence, TINA is the acronym for the famous Thatcherism "there is no alternative."

CHAPTER 5

Chapter title. See Rodríguez Kuri (2010) for similar usage. Here as in quotidian contemporary parlance, *city-state* makes reference to the Federal District's complicated territorial and legal relationships with the Mexican subnational unit of the estado.

1. Elvira Vargas and Romero Sánchez (2016).

2. *Manifestación* is the word nearly always used to describe public protests, marches, or other large planned or unplanned public gatherings of a political nature.

3. The Museum of Mexico City, located on Isabel la Católica street in the city's historic center, is a converted mansion, once the familial home of the Counts of Santiago de Calimaya gifted by Hernán Cortés to loyal conquistador Juan Gutiérrez Altamirano.

4. The word typically used was *truco*, variously translated as "trick," "ploy," or "ruse."

5. See Monsiváis (1997); Billig (2005); Neria and Aspinwall (2015); Van Ramshorst (2019); Gerlofs (2022).

6. Ebrard had felt the sting of federal control most personally and most harshly even before his tenure as mayor. In 2004, President Vicente Fox exercised his federal privilege in dismissing Ebrard from his post as secretary of public security under AMLO's mayoralty, citing Ebrard's poor handling of the lynching of two federal police officers two weeks prior in the Tláhuac borough, in the far southeastern part of the Federal District (Elvira Vargas 2004). The ability of the president of the Republic to dismiss political appointees of the mayor of the Federal District would later become a major part of the campaign for additional political reforms during Ebrard's sexenio.

7. Even without such provision, the reforms faced a considerable measure of opposition in the legislature.

8. See the analysis and commentary of Cárdenas Gracia (2017) and Rabell García (2017), among others.

9. *Constituyente* became an important word in 2016, though there was enough confusion surrounding its meaning that some of the city's promotional materials for the process included explanatory details for its proper use. In the singular, it can refer to the constitutional process as a whole (from the reforms of January 2016 though the adoption of the city's constitution in January 2017), the Constitutional Assembly as a body (here denoted with an initial capital, Constituyente), or to individual members of the Constitutional Assembly in particular.

10. Enrique Ortiz Flores, an important member of the CPCCMDC and still a leading civil society figure, was one of Mancera's twenty-eight notables.

11. The city claims to have held fifty-five official meetings of this type in 2016, which can be seen through a scrollable calendar at: http://www.constitucion.cdmx.gob.mx/participa/#dialogos-publicos (last accessed June 12, 2017).

12. Both are major political figures in the city and longtime PRD (and later Morena, in the case of Muñoz Ledo) members. Encinas served briefly as the city's jefe de gobierno (2005–2006) when AMLO left the office to run for the national presidency, and is currently serving as AMLO's Subsecretario de Derechos Humanos, Migración y Población de México. He left the PRD in January of

2015, citing corruption of the party in the wake of the forty-three missing students in the town of Iguala (in whose disappearance the town's PRD mayor was implicated) (Muñoz 2015a). Muñoz Ledo was a founding member of the PRD and former president of the both the PRI (1975–1976) and PRD (1993–1996), and has served in numerous government posts since the early 1970s, including secretary of labor, secretary of energy, and ambassador to the European Union. He began a run for the national presidency in the election of 2000 as a candidate of the now defunct *Partido Auténtico de la Revolución Mexicana* (Authentic Party of the Mexican Revolution, hereafter PARM), but resigned in support of eventual PAN victor Vicente Fox. In September, 2020, he was one of some twenty political figures vying for the presidency of Morena.

13. The participation process also reportedly included some 602 citizen proposals submitted to the IEDF and the opportunity to read and make comments on some 30 essays submitted by notables to the online document sharing platform pubpub.org, according to the city's official constitutional webpage at: http://www.constitucion.cdmx.gob.mx/

14. The notables did have a plan in place to deal with the most formalized part of this process, the proposals submitted through Change.org. The crafters of the top proposals, measured in votes of support from the general public that numbered over ten thousand, were invited to present their ideas to a smaller working group, who would relay the proposals to the larger group of drafters. Proposals receiving over fifty thousand votes were promised an invitation to relate their ideas personally to the entire group of notables (Campoy 2016).

15. Some international observers gushed with praise over the ultimate document and the process of its construction. Citing its stance on often marginalized groups, Weiss (2017) argued that the constitution "reads like a progressive manifesto," and the Progressive Alliance (2016) called it "among the most advanced in Latin America and the world."

16. See Ackerman (2016), among others, as exemplary of this tendency. Notables, Mancera repeatedly reassured the public, were chosen not based on party affiliation but rather on their being "people who know the city, who have worked for the city or who are experts in their field" (Durán 2016). Among the notables, he told *Proceso* in February of 2016, "We have good coverage of themes, from indigeneity, culture, the feminist movement to the UNAM Institute of Juridical Investigations, and the same from the National Polytechnic [University]" (Durán 2016). Interestingly, as with other instances discussed in Chapter 4 and elsewhere, Mancera's supposed ability to trade influence on this council is here legitimated by a rhetoric of transparently shifting away from nepotism and party loyalty. On this reading, purported democratization provides a veil that makes it possible to conduct traditional political horse trading.

17. Though they ultimately came to naught, various rumors swirling around the Peña Nieto/Mancera alliance during my fieldwork from 2015–2017 had it that

Mancera's 2018 candidacy would receive some type of support from the PRI, either in the form of a PRI candidacy for Mancera (however unlikely), a combined PRI/PRD ticket (also highly unlikely), some type of tacit support via political machinery, or at the very least an unobstructionist stance.

18. See especially Ziccardi (2017, translation by author), who argued on the heels of Peña Nieto's reforms for the necessity of "institutional redesign" at the metropolitan scale, in service to coordinated, multilevel urban governance. See also Fernandes (2007, 211), who gives the name "the metropolitan question," to this intractable imperative of regional "intergovernmental articulation" in their study of the right to the city in Brazil.

19. According to INEGI and as shown in Figure 5.9 (according to 2010 INEGI figures), the 58 municipios of the Estado de México that belong to the ZMCM were home to 11,529,701 of that state's 15,175,862 residents, or 75.97 percent (INEGI 2017). It's also worth noting that residents of México and Hidalgo make up a substantial portion of the population of the metropolitan area overall. Also according to 2010 INEGI figures, Ziccardi (2014, 210) reports that "over half (53 per cent) of the [ZMVM] population lives in surrounding municipalities of Mexico State, 42 per cent in the Federal District, and 5 per cent in Hidalgo State municipalities."

20. Ward (1989) explained that even by the late 1980s the city's problems and the immense chore that urban governance had become were beyond the bounds of a single governing entity, but, though the idea had been discussed for some time already, it had received little political traction. This, Ward argues, was a symptom of the PRI's approach to governance. "In Mexico City," he wrote, "the power structure that has evolved is one of control, not one of development" (Ward 1989, 320).

21. This vote was to decide proportional party-representation on the Assembly, though there were also independent candidates. Proportional representation was divided based on the proportion of votes each party received out of the valid votes, or the total number of votes minus the null votes. In the final tally, the valid votes numbered 1,926,608. The total number of votes (2,092,721) constituted a participation rate of 28.6791 percent (INE 2016). The results of the vote yielded the following number of seats for each party ultimately represented: Morena (22); PRD (19); PAN (7); PRI (5); Partido Encuentro Social (2); Nueva Alianza (2); Movimiento Ciudadano (1); Partido Verde Ecologista de México (1). Ismael Figueroa Flores, head of Mexico City's firefighters union, was the only independent candidate to win a seat on the Assembly. Though initially some ten thousand votes short of the roughly thirty-two thousand necessary for an independent to be seated, Figueroa was given a seat that was left vacant after the other proportional seats had been divided (given the inability to award a fraction of a seat to a party), by virtue of his having received more votes as an independent than had the highest vote-getting unrepresented party (*El Financiero* 2016).

22. Within weeks of the Assembly's decision, the document and its proponents faced several court challenges emanating from the national political sphere. These cases charged various encroachments upon national authority in the former Federal District, and sought to prevent the constitution's acquisition of force. Alejandro Encinas Rodríguez, who had been chosen by the Assembly to serve as its President in October of 2016 (see Llanos Samaniego 2016), threatened in February of 2017 to reconvene the Assembly in order to carefully reword the offending sections and nullify the most serious of challenges, which had already reached the country's Supreme Court (see Nuñez 2017).

23. These foundations similarly make up the second chapter of the Charter, as discussed in Chapter 3.

24. The presence of the right to the city, I was also told, was a point of some contention among the notables, for some of whom it represented a political moment that had passed, and for others of whom it was merely the particular concern of an older generation of political figures. In either case, it was considered old hat by some, though I never heard this stated openly in public.

25. Except for the citizen initiative, which evidently cannot achieve this status. The citizen initiative subsection (B) of Article 25 explains that citizens may develop proposals for laws or even amendments to this very same constitution, and that these will be considered by the city's Congress (rather than having to funnel such requests through the executive branch) if they receive signatures totaling at least 0.3 percent of the city's electorate.

26. Including by such figures as former jefe and current (as of September 2020) Mexican foreign minister Marcelo Ebrard Casaubón.

27. See Mitchell (2003).

28. As discussed in Chapter 4, this rhetorical adjustment reflects the party's highly consequential relationship with processes of democratization. Another notable attempt to steal away the Revolutionary mantel was made by the PARM, an alternative party whose claim to authenticity was never rewarded with much electoral success.

29. See Gutmann (2002).

30. As Williams (2002, 98) instructively reminds, "there are in fact no masses, but only ways of seeing people as masses. With the coming of industrialism, much of the old social organization broke down and it became a matter of difficult personal experience that we were constantly seeing people we did not know, and it was tempting to mass them, as 'the others,' in our minds. Again, people were physically massed, in the industrial towns, and a new class structure (the names of our social classes, and the word 'class' itself in this sense, date only from the Industrial Revolution) was practically imposed."

31. As noted above, the PRD has also fallen victim to several notable defections from within its leadership, including Cuauhtémoc Cárdenas in 2014 (see Muñoz and Saldierna 2014), Alejandro Encinas in 2015 (see Muñoz 2015a), then party leader Porfirio Muñoz Ledo in 2000 (see Martínez and Pérez 2000),

Rosario Robles in 2012 (see *Sin Embargo* 2012), Marcelo Ebrard in 2015 (see Muñoz 2015b) and obviously AMLO himself in 2012. Some of these figures continue to work closely with the PRD, while others, like AMLO, have been openly critical and even hostile toward their former party. Others joined competing parties, like Robles, who was appointed to Peña Nieto's cabinet as secretary of social development in 2012. Ebrard's exit, wherein he cited party leadership's creeping proximity to Peña Nieto, completed the set of the five PRD mayors, save for the then-sitting Mancera who alone remained with the party. See Ross (2008) for an especially colorful telling of the demise of the PRD amid its factional strife during the PAN presidencies in particular.

32. Using the abbreviation Morena (or MoReNa) for the party's full name, Movimiento de Regeneración Nacional (The National Regeneration Movement) intentionally plays on this overcoded term (see Chapter 2) in an assertive way, foregoing the standard abbreviation of the other major parties (Morena would hypothetically be MRN).

33. A particularly tense debate was set off by Medium's publication of Tamara Velasquez's (2017) response to VICE. Velasquez maintains that despite its glossy veneer, Mexico City's poverty and inequality fly in the face of any attempt to christen it the new Berlin.

34. See Campoy (2016); Melchor (2016); Flores (2017); Progressive Alliance (2017); and Scruggs (2017).

35. In a thoroughgoing treatment of a foundational tension over this concept between Adolfo Sánchez Vázquez and Bolívar Echeverría, for instance, Gandler (2016, 100) forcefully asserts, "Therefore, in Marx's concept of praxis the immense achievements on the theoretical terrain of both previous materialism and idealism find their place; following Sánchez Vázquez, we could boldly ask if this concept does not simultaneously transcend, maintain, and suspend (in the Hegelian sense of 'aufheben') the entire dichotomy between materialism and idealism."

36. On this tendency of rights talk, see especially Brown (2004); Žižek (2005); Hodgson (2011). On the relationship between industrialization and urbanization according to Lefebvre, see Lefebvre (2003a); Harvey (1973); Castells (1977).

37. See Lomnitz (2005) for a compelling overview.

38. There is strong resonance here also with Brown's (2011) invitation to rethink democracy itself as a perpetually unfinished project, as exemplified in the epigraph to this chapter.

39. See Tamayo (2015, 177, my translation), for whom later urban movements can be thought of as "the true inheritance of the movement begun in 1968."

40. For prominent examples from this cultural moment, see José Vasconcelos Calderón's (1997) notion of Mexican mestizos as the cosmic race, and Diego Rivera's famous 1948 mural, *Sueño de una Tarde Dominical en la Alameda Central* (Dream of a Sunday afternoon in Alameda park).

41. This point is perhaps best illustrated by a chant that commonly interrupts public meetings and political gatherings in the city, wherein one or more participants shout "Zapata vive!" (Zapata lives!) to which a mass of others respond "La lucha sigue!" (The struggle continues!). Emiliano Zapata, the revolutionary leader of dispossessed peasants whose deep distrust of the capital city and his adherence to rural aesthetic norms are the stuff of legend, now inhabits a favorite rallying cry of a significant segment of el monstruo's grassroots urban Left.

42. This section title is inspired by the 2010 essay "A Future for Mexico" authored by former Foreign Affairs Minister Jorge Castañeda and author and historian Hector Aguilar Camin, the introduction to which is entitled, "The Weight of History." Castañeda and Aguilar claim that "Mexico is a prisoner of its history," and that it "needs to be emancipated from its past." Theirs is a perspective that emphasizes the failures of the last century of Mexican politics (even the so-called transition to democracy) and the need for a kind of radical break with the norms established during the PRI's reign.

CONCLUSION

1. The October 2018 national consulta, which solicited public affirmation or negation of the Nuevo Aeropuerto Internacional de México (NAICM), which reportedly gave AMLO the political mandate and cover for this action, was run not by the INE but by Morena itself, causing PRI, PRD, and PAN leaders, among others, to abstain in protest. Others, notably including Swiss bank UBS, also decried what they considered presidential overreach, with AMLO exercising the powers of the Presidency before his inauguration. See *La Jornada* (2018); Jiménez and Garduño (2018).

      Línea 12 has been and continues to be a source of controversy in local and national politics, especially with respect to its initial financing and construction, its structural assessment (or lack thereof) in the wake of the 2018 earthquake and years of citizen demands for investigations, and the discomforting (for many) relationships between individuals, parties, and firms at the center of these controversies from the line's inception (perhaps especially Marcelo Ebrard, AMLO, Miguel Ángel Mancera, Enrique Peña Nieto, and Carlos Slim). The collapse took place on May 3, 2021 in the colonia of Tláhuac, and killed some twenty-six persons. See *BBC News Mundo* (2021) and Kitroeff et al. (2021) for useful overviews.

      AMLO's public appeals to faith in the military, as opposed to a corrupt Policía Federal, were an important part of the justification for a Guardia Nacional. See Melimopoulos (2019); Ahmed (2020). With pressure reportedly applied from the highest levels of the Mexican state (notably including from AMLO and Marcelo Ebrard), the charges against Cienfuegos were dropped in November 2020 (Sieff and Jacobs 2020).

The June 2021 elections provide an especially striking illustration of these challenges. The vote essentially divided the city east and west, with a PRI/PRD/PAN alliance (organized as the *fuerza política* known as Va por México) garnering a majority for *alcalde* (executive head of each of the city's sixteen alcaldías), for instance, in the nine alcaldías of Azcapotzalco, Coyoacán, Cuajimalpa de Morelos, La Magdalena Contreras, Álvaro Obregón, Tlalpan, Benito Juárez, Cuauhtémoc, and Miguel Hidalgo, and Morena (in coalition with the PT and PVEM as Juntos Hacemos Historia) taking the seven alcaldías of Gustavo A. Madero, Iztacalco Iztapalapa, Milpa Alta, Tláhuac, Venustiano Carranza, and Xochimilco. This stands in contrast to the electoral cartography of the 2018 elections, in which the then-ascendent Morena coalition (also Juntos Hacemos Historia, with the PT and Encuentro Social) won the eleven alcadías of Azcapotzalco, Gustavo A. Madero, Iztacalco, Iztapalapa, La Magdalena Contreras, Álvaro Obregón, Tláhuac, Tlalpan, Xochimilco, Cuauhtémoc, and Miguel Hidalgo, while a PAN/PRD alliance (along with Movimiento Ciudadano, as the coalition, Por la Ciudad de México al Frente) won only the four alcaldías of Coyoacán, Milpa Alta, Benito Juárez, and Venustiano Carranza, and the PRI won only Cuajimalpa de Morelos, per IECM (2018, 2021).

2. See Latour (1987); Mitchell (1996).

3. See also Pickvance (2003), who labels this Castells's restrictive use of the term *urban social movement*, as opposed to the generic use, which left open the possibility of qualification simply on the merits of lofty goals, regardless of outcomes.

4. See especially Young (1990); Fraser (1997); Faintstein (2010); Sharp (2016); Yusoff (2018).

APPENDIX

1. Evident in the dialectical use of common terms like *context, tension, influence,* and *determination.*

2. To mention only a few, my approach owes a great deal, in various ways, to the work of Diane Davis, Susan Eckstein, Mustafa Dikeç, Daniel Goldstein, James Holston, and Asef Bayat.

3. It was in some of these meetings that I first began to become aware of being awarded special attention and/or privileges. Much to my surprise, I was sometimes mistaken for a person of some importance or influence, which is likely attributable to some combination of my attire (often rather formal), conduct (sometimes speaking before, during, or after meetings with officials, assistants, or social movement leaders or NGO personnel, and usually—though not always—taking furious notes during meetings), and position in the microgeographies of these events (I was often offered a good seat at meetings, or arrived early enough to find a good place to stand).

4. Ignacio also accompanied me to countless events across the city, especially on weekends. He took to calling these outings *aventuras* (adventures)—a word

often used to cheekily describe extramarital affairs—both because he considered both the locations (from museums and archives to markets, altars, and the site of Iztapalapa's world famous Via Crucis) and nature of these outings (many were academic or cultural in nature) so far afield of his routine as to seem adventurous, and because he often described, introduced, or tagged me in photos as his husband during or after these excursions (a practice that continues to provoke confusion from others and seems to produce endless amusement for him). I am inestimably grateful for his contributions.

5. My understanding and practice of these modes of observation and recording owe much to the work of Loïc Wacquant, and to inspiring and instructive remarks of and conversations with Asher Ghertner, Nina Siulc, Rossana Reguillo, Deborah Pellow, Gretchen Purser, Rosemary-Claire Collard, and Rafael Sánchez, among others.

6. Rutgers University (2012–2017); UNAM (2015–2016); Dartmouth College (2017–2018); Princeton University (2018–2020).

7. I was inevitably forced to make some choices within such an impossibly large landscape. Early in my fieldwork, I decided to focus more attention on those events and topics that were receiving more attention, though I also routinely surveyed particular topics and discussions as the contours of the book emerged (such as the right to the city, the anti-Corredor movement, and the city's constitutional process), or out of particular interest (such as the city's *parquimetros*, the poisoning of some eighteen dogs in Parque México, and the city's air quality).

References

Ackerman, John. 2012. "The Return of the Mexican Dinosaur." *Foreign Policy*, July 2, 2012.

———. 2014. "Mexico: The Myth of a Democratic Transition." *Proceso*, May 18, 2014.

———. 2016. "Emancipación Capitalina." *La Jornada*, March 28, 2016.

Adler, David. 2015(a). "Do We Have a Right to the City?" *Jacobin*, October 6, 2015(a).

———. 2015(b). "El Derecho a la Ciudad, un Sueño Incumplido en el DF." *Nexos*, November 10, 2015(b).

ADNPolítico. 2012. "La Reforma Política que Propone el 'Pacto por México.'" *ADNPolítico*, December 2, 2012.

Aguirre, Javier. 1999. "Constitutional Shift Toward Democracy: Mexico City's Amendment to Grant Elections Gives Rise to a New Constitutional Order." *Loyola of Los Angeles International and Comparative Law Review* 21, no. 1: 131–57.

Ahmed, Azam. 2020. "Mexico's Former Defense Minister Is Arrested in Los Angeles." *New York Times*, October 16, 2020.

Ai Camp, Roderic. 2000. "The Time of the Technocrats and Deconstruction of the Revolution." In *The Oxford History of Mexico*, edited by William H. Beezley and Michael C. Meyer, 609–36. New York: Oxford University Press.

———. 2014. *Politics in Mexico: Democratic Consolidation or Decline?* 6th ed. New York: Oxford University Press.

Alarcón, Diana, and Terry McKinley. 1998. "Increasing Wage Inequality and Trade Liberalization in Mexico." In *Economic Reform and Income Distribution in Latin America*, edited by Albert Berry, 138–53. Boulder, CO: Lynne Rienner Publishers.

Alexander, Ryan. 2016. "Mexican Politics, Economy, and Society, 1946–1982." *Oxford Research Encyclopedia of Latin American*

History. http://latinamericanhistory.oxfordre.com/view/10.1093/
acrefore/9780199366439.001.0001/acrefore-9780199366439-e-261?print=pdf.

Allmendinger, Phil, and Graham Haughton. 2012. "Post-Political Spatial Planning
in England: A Crisis of Consensus?" *Transactions of the Institute of British
Geographers* 37, no. 1: 89–103.

Álvarez Bejar, Alejandro. 2016. "Tribute to Alonso Aguilar Monteverde: Ten
Key Policies for Understanding the Neoliberal Transformation of Mexican
Capitalism." *Social Justice* 42, no. 1:107–15.

Angeles-Castro, Gerardo. 2011. "Economic Liberalisation and Income
Distribution: Theory and Evidence in Mexico." In *Market Liberalism, Growth,
and Economic Development in Latin America*, edited by Gerardo Angeles-
Castro, Ignacio Perrotini-Hernández, and Humberto Ríos-Bolívar, 195–219.
London: Routledge.

Arendt, Hannah. (1963) 2006. *On Revolution*. Reprint, New York: Penguin.

Asamblea Legislativa del Distrito Federal (ALDF). "Denuncian Falta de
Mantenimiento en Parque Cuitláhuac." Accessed May 25, 2017. http://
www.aldf.gob.mx/comsoc-denuncian-falta-mantenimiento-parque-
cuitlahuac--13268.html.

Attoh, Kafui. 2011. "What Kind of Right Is the Right to the City?" *Progress in
Human Geography* 35, no. 5: 669–85.

Auyero, Javier. 1999. "'This Is a Lot Like the Bronx, Isn't It?': Lived Experiences
of Marginality in an Argentine Slum." *International Journal of Urban and
Regional Research* 23, no. 1: 45–69.

Ayrelan Iuvino, Amelia. 2017. "The Speakeasies of Mexico City." *GQ*, May 1, 2017.

Azuela, Antonio. 2007. "Mexico City: The City and Its Law in Eight Episodes,
1940–2005." In *Law and the City*, edited by Andreas Philippopoulos-
Mihalopoulos, 153–69. New York: Routledge.

Ball, Michael, and Priscilla Connelly. 1987. "Capital Accumulation in the Mexican
Construction Industry, 1930–1982." *International Journal of Urban and
Regional Research*, 11, no. 2: 153–71.

Barros, Cristina. 2010. "Supervía Poniente." *La Jornada*, May 23, 2010.

Baverstock, Alasdair. 2014. "Miserable Flooding Is a Fact of Life in Mexico City's
Impoverished Borough." *VICE*, June 2, 2014.

Bayat, Asef. 2013. *Life as Politics: How Ordinary People Change the Middle East*.
2nd ed. Redwood City, CA: Stanford University Press.

BBC News Mundo. 2021. "Accidente el Línea 12: La Controvertida Historia de
la Línea de Metro que Colapsó y Provocó Decenas de Muertos y Heridos."
BBC News Mundo, May 4, 2021. https://bbc.com/mundo/noticias-america-
latina-56980066.

Beauregard, Luis Pablo. 2016. "La Primera Gran Constitución del Siglo XXI Será
la de la Ciudad de México." *El País*, March 24, 2016.

Beauregard, Robert. 2018. *Cities in the Urban Age: A Dissent*. Chicago: University
of Chicago Press.

Becerra Chávez, Pablo Javier. 2003. "Las Reformas Electorales en la Transición Mexicana." In *Contexto y Propuestas para una Agenda de Reforma Electoral en México*, edited by Pablo Javier Becerra, Víctor Alarcón Olguín, Cuitláhuac Bardán Esquivel, 15–23. Ciudad de México: Instituto de Investigaciones Legislativas del Senado de la República. https://archivos.juridicas.unam.mx/www/bjv/libros/7/3182/5.pdf.

Becerril, Andrea, Georgina Saldierna, Roberto Garduña, and Enrique Mendez. 2006. "AMLO, 'Presidente Legítimo'; Toma Posesión el 20 de Noviembre: CND." *La Jornada*, September 17, 2006.

Becker, Anne, and Markus-Michael Müller. 2013. "The Securitization of Urban Space and the "Rescue" of Downtown Mexico City." *Latin American Perspectives* 189, no. 2: 77–94.

Beezley, William. (1987) 2004. *Judas at the Jockey Club and Other Episodes of Porfirian Mexico*. 2nd ed. Lincoln: University of Nebraska Press.

Benería, Lourdes. 1989. "Subcontracting and Employment Dynamics in Mexico City." In *The Informal Economy: Studies in Advanced and Less Developed Countries*, edited by Alejandro Portes, Manuel Castells, and Lauren A. Benton, 173–88. Baltimore, MD: Johns Hopkins University Press.

Benjamin, Walter. (1968) 2007. *Illuminations*. Reprint, New York: Schocken Books.

Bergman, Lowell. n.d. "The Hank Family of Mexico: Background and an Interview with Carlos Hank Rhon." *Frontline*, accessed October 17, 2022. http://www.pbs.org/wgbh/pages/frontline/shows/drugs/special/hankfam.html.

Beyer, Glenn H. 1967. *The Urban Explosion in Latin America: A Continent in the Process of Modernization*. Ithaca: Cornell University Press.

Billig, Michael. 2005. *Laughter and Ridicule: Towards a Social Critique of Humor*. London: Sage.

Blanco, José Joaquín. 2004. "Las Lomas I." In *The Mexico City Reader*, edited by Rubén Gallo, 108–113. Madison: University of Wisconsin Press.

———. 2004. "Cuauhtémoc." In *The Mexico City Reader*, edited by Rubén Gallo, 222–26. Madison: University of Wisconsin Press.

Blokland, Talja, Christine Henschel, Andrej Holm, Henrik Lebuhn, and Talia Margalit. 2015. "Urban Citizenship and the Right to the City: The Fragmentation of Claims." *International Journal of Urban and Regional Research* 39, no. 4: 655–65.

Bolaño, Roberto. (1998) 2007. *The Savage Detectives*. Reprint, Picador, New York.

Borja, Jordi, Davinder Lamba, Enrique Ortíz Flores, Leticia Marques Osorio, Asier Martínez, Gerardo Pisarello Prados, Nelson Saulé Junior, Joseph Schechla, Sebastián Tedeschi, and Favio Velázquez. 2006. *The Right to the City around the World*. Mexico City: Habitat International Coalition-América Latina. hlrn.org/igm/documents/The%20Right%20Right%20to%20the%20City%20around%20the%20World.pdf

Boudreau, Julie-Anne. 2017. *Global Urban Politics: Informalization of the State*. Cambridge: Polity Press.

———, and Felipe de Alba. 2020. "*El Monstruo*: Reflections on Catastrophic Metaphors about Mexico City." In *Handbook of Megacities and Megacity-Regions*, edited by Danielle Labbé and André Sorensen, 65–76. Cheltenham: Edward Elgar.

Brenner, Neil. 2000. "The Urban Question as a Scale Question: Reflections on Henri Lefebvre, Urban Theory and the Politics of Scale." *International Journal of Urban and Regional Research* 24, no. 2: 361–78.

———, and Christian Schmid. 2014. "The 'Urban Age' in Question." *International Journal of Urban and Regional Research* 38, no. 3: 731–55.

Brown, Alison. 2013. "The Right to the City: Road to Rio 2010." *International Journal of Urban and Regional Research* 37, no. 3: 957–71.

———, and Annali Kristiansen. "Urban Policies and the Right to the City: Rights, Responsibilities, and Citizenship." UNESCO. http://unesdoc.unesco.org/images/0017/001780/178090e.pdf.

Brown, Wendy. 2004. "'The Most We Can Hope For': Human Rights and the Politics of Fatalism." *South Atlantic Quarterly* 103, no. 2/3: 451–63.

———. 2011. "'We Are All Democrats Now…'" In *Democracy in What State?*, by Giorgio Agamben, Alain Badiou, Daniel Bensaïd, Wendy Brown, Jean-Luc Nancy, Jacques Rancière, Kristin Ross, and Slavoj Žižek, 44–57. New York: Columbia University Press.

Bruhn, Kathleen. 1997. *Taking on Goliath: The Emergence of a New Left Party and the Struggle for Democracy in Mexico*. University Park: Penn State University Press.

———. 2008. *Urban Protest in Mexico and Brazil*. Cambridge: Cambridge University Press.

Buckley, Michelle, and Kendra Strauss. 2016. "With, Against and Beyond Lefebvre: Planetary Urbanization and Epistemic Plurality." *Environment and Planning D: Society and Space* 34, no. 4: 617–36.

Buechler, Steven. 1995. "New Social Movement Theories." *Sociological Quarterly* 36, no. 3: 441–64.

Caistor, Nick. 2000. *Mexico City: A Cultural and Literary Companion*. Oxford: Signal Books.

Caldeira, Teresa. 2017. "Peripheral Urbanization: Autoconstruction, Transversal Logics, and Politics in Cities of the Global South." *Environment and Planning D: Society and Space* 35, no. 1: 3–20.

Camacho de Schmidt, Aurora, and Arthur Schmidt. 1995. "Foreword: The Shaking of a Nation." In *Nothing, Nobody: The Voices of the Mexico City Earthquake*, by Elena Poniatowska, ix–xxix. Philadelphia: Temple University Press.

Campos, Raymundo, Gerardo Esquivel, and Nora Lustig. 2012. "The Rise and Fall of Income Inequality in Mexico, 1989–2010." Rome: Society for the Study of Economic Inequality, Working Paper 267, September 2012.

Campoy, Ana. 2016. "Mexico City Is Crowdsourcing its New Constitution Using Change.org in a Democracy Experiment." Quartz, April 21, 2016. https://

qz.com/662159/mexico-city-is-crowdsourcing-its-new-constitution-using-change-org-in-a-democracy-experiment.

Candiani, Vera. 2014. *Dreaming of Dry Land: Environmental Transformation in Colonial Mexico City.* Redwood City, CA: Stanford University Press.

Cantú, Francisco. 2019. "The Fingerprints of Fraud: Evidence from Mexico's 1988 Presidential Election." *American Political Science Review* 113, no. 3: 710–26.

Cárdenas Gracia, Jaime Fernando. 2017. "Comentario sobre la Reforma Política de la Ciudad de México." *Cuestiones Constitucionales* 36, (January–June): 367–94.

Cardoso, Fernando Henrique, and Enzo Falleto. (1971) 1979. *Dependency and Development in Latin America.* Reprint, Berkeley: University of California Press.

Del Carmen Moreno Carranco, Maria. 2008. "The Socio/Spatial Production of the Global: Mexico City Reinvented through the Santa Fe Urban Megaproject." PhD diss., University of California at Berkeley.

Casey, Nicholas. 2011. "Emperor's New Museum." *Wall Street Journal*, March 3, 2011.

Cason, Jim, and David Brooks. 1998. "Desmienten SRE y Gobernación a Diario Estadounidense que Liga a Labastida con el Narco." *La Jornada*, February 6, 1998.

Castañeda, Jorge G., and Héctor Aguilar Camín. 2009. "Un Futuro para México." *Nexos*, November 1, 2009.

Castells, Manuel. (1972) 1977. *The Urban Question: A Marxist Approach.* Reprint, Cambridge, MA: MIT Press.

———. 1983. *The City and the Grassroots: A Cross-Cultural Theory of Urban Social Movements.* Berkeley: University of California Press.

Castillo García, Gustavo. 2001. "A Siete Años del Asesinato de José Francisco Ruiz Massieu, el Expediente Continúa Abierto." *La Jornada*, September 28, 2001.

Castillo García, Gustavo. 2020. "Ratifica Juez Permanencia de Rosario Robles en Prisión Preventiva." *La Jornada*, May 1, 2020.

Castree, Noel. 1996. "Birds, Mice and Geography: Marxisms and Dialectics." *Transactions of the Institute of British Geographers* 21, no. 2: 342–62.

Celorio, Gonzalo. 2004. "Mexico: City of Paper." In *The Mexico City Reader*, edited by Rubén Gallo, 33–52. Madison: University of Wisconsin Press.

Centeno, Miguel. (1994) 1997. *Democracy within Reason: TecÚocratic Revolution in Mexico.* Reprint, University Park: Penn State University Press.

Chabat, Jorge. 2002. "Mexico's War on Drugs: No Margin for Maneuver." *Annals of the American Academy of Political and Social Science* 582 (July): 134–48.

Clark, Patrick. 2015. "To Hike Rent, Landlords Swap Plain Walls for Exposed Brick." *Bloomberg*, January 29, 2015.

Código. 2015. "Corredor Cultural Chapultepec: 6 Expertos Opinan." *Código*, August 26, 2015.

Comisión de Derechos Humanos del Distrito Federal (CDHDF). 2011. *Recomendación 1/2011: Report CDHDF/III/122/AO/10/D4636.* Mexico City: Comisión de Derechos Humanos del Distrito Federal (CDHDF).

Comité Promotor de la Carta de la Ciudad de México por el Derecho a la Ciudad (CPCCMDC). 2011. *La Carta de la Ciudad de México por el Derecho a la Ciudad*. Mexico City, Mexico: Comité Promotor de la Carta de la Ciudad de México por el Derecho a la Ciudad (CPCCMDC).

Connelly, Priscilla. 1988. "Crecimiento Urbano, Densidad de Población y Mercado Inmobiliario." *Revista A* 9, no. 25: 61–85.

———. *The Case of Mexico City, Mexico*. UN-Habitat. http://www.ucl.ac.uk/dpu-projects/Global_Report/pdfs/Mexico.pdf.

Corcoran, Steven. 2010. "Editor's Introduction." In *Dissensus: on Politics and Aesthetics*, edited and translated by Steven Corcoran, 1–31. New York: Bloomsbury.

Cox, Kevin, and Andrew Mair. 1989. "Levels of Abstraction in Locality Studies." *Antipode* 21, no. 2: 121–32.

Crane, Nicholas. 2015. "Politics Squeezed through a Police State: Policing and *Vinculación* in Post-1968 Mexico City." *Political Geography* 47 (June): 1–10.

Crouch, Colin. 2004. *Post-Democracy*. Malden: Polity Press.

Crossa, Veronica. 2009. "Resisting the Entrepreneurial City: Street Vendors' Struggle in Mexico City's Historic Center." *International Journal of Urban and Regional Research* 33, no. 1: 43–63.

———. 2013. "Play for Protest, Protest for Play: Artisan and Vendors' Resistance to Displacement in Mexico City." *Antipode* 45, no. 4: 826–43.

———. 2016. "Reading for Difference on the Street: "De-homogenizing Street Vending in Mexico City." *Urban Studies* 53, no. 2: 287–301.

Cruikshank, Barbara. 1999. *The Will to Empower: Democratic Citizens and other Subjects*. Ithaca: Cornell University Press.

Dagnino, Evelina. 2007. "Citizenship: A Perverse Confluence." *Development in Practice* 17, no. 4–5: 549–56.

Dávalos, Renato. 2004. "Revela Rosario Robles que su Renuncia Atendió a una Petición de Leonel Godoy." *La Jornada*, March 11, 2004.

Davidson, Mark, and Kurt Iveson. 2015. "Recovering the Politics of the City: From the 'Post-political' City to a 'Method of Equality' for Critical Urban Geography." *Progress in Human Geography* 39, no. 5: 543–59.

Davis, Diane E. 1994. *Urban Leviathan: Mexico City in the Twentieth Century*. Philadelphia: Temple University Press.

———. 2009. "The Modern City: From the Reforma-Peralvillo to the Torre Bicentenario: The Clash of 'History' and 'Progress' in the Urban Development of Modern Mexico City." In *Mexico City through History and Culture*, edited by Linda Newson and John King, 55–82. New York: Oxford University Press.

———. 2014. "Competing Globalizations in Mexico City's Historic Centre." *Scapegoat* 6, no. 1: 155–66.

Davis, Diane E., and Christina Rosan. 2004. "Social Movements in the Mexico City Airport Controversy: Globalization, Democracy, and the Power of Distance." *Mobilization: An International Journal* 9, no. 3: 279–93.

Delgadillo, Victor. 2016. "Selective Modernization of Mexico City and its Historic Center. Gentrification without Displacement?" *Urban Geography* 37, no. 8: 1154–74.

DePalma, Anthony. 1994(a). "Top Presidential Candidate Is Assassinated in Mexico." *New York Times*, March 24, 1994(a).

———. 1994(b). "A Year to Forget: 1994 Leaves Mexico Reeling." *New York Times*, December 29, 1994(b).

———. 1995. "Mexico's 'Accidental' Chief Spurns Role of Strongmen." *New York Times*, December 28, 1995.

Derickson, Kate. 2017. "Taking Account of the 'Part of Those that Have No Part.'" *Urban Studies* 54, no. 1: 44–48.

Dfensor. 2016. "Voces Ciudadanas en la Constitución CDMX." *Dfensor* 14, no. 7 (2016): 54–57.

Dikeç, Mustafa. 2005. "Space, Politics, and the Political." *Environment and Planning D: Society and Space* 23, no. 2: 171–88.

———. 2007. *Badlands of the Republic: Space, Politics and Urban Policy.* Malden: Blackwell Publishing.

———. 2013. "Beginners and Equals: Political Subjectivity in Arendt and Rancière." *Transactions of the Institute of British Geographers* 38, no. 1: 78–90.

———. 2017. "Disruptive Politics." *Urban Studies* 54, no. 1: 49–54.

———, and Erik Swyngedouw. 2017. "Theorizing the Politicizing City." *International Journal of Urban and Regional Research* 41, no. 1: 1–18.

Dixon, Deborah, Keith Woodward, and John Paul Jones. 2008. "Guest Editorial: On the Other Hand . . . Dialectics." *Environment and Planning A* 40, no. 11: 2549–61.

Dolan, Kerry. 2011. "The World's Richest Man's Favorite Architect." *Forbes*, April 15, 2011.

Dresser, Denise. 2014. "El Pasado Omnipresente." *Proceso*, October 12, 2014.

Duhau, Emilio. 2014. "The Informal City: An Enduring Slum or a Progressive Habitat?" In *Cities from Scratch: Poverty and Informality in Urban Latin America*, edited by Brodwyn Fischer, Bryan MCann, and Javier Auyero, 150–69. Durham, NC: Duke University Press.

Durán, M. 2016. "Inicia Proceso Constituyente." *Reforma*, February 23, 2016.

Ebrard Casaubón, Marcelo. "Marcelo Ebrard y el Derecho a la Ciudad Campaña," July 31, 2008. Habitarte, YouTube, posted August 7, 2008, 2:28. https://www.youtube.com/watch?v=DfaOgb5M-GU.

———. "Palabras de Marcelo Ebrard en la Firma de la Carta por el Derecho a la Ciudad." Noeve Sietesoles, YouTube, July 13, 2010, 4:14. https://www.youtube.com/watch?v=leBUSbWtPrw.

Echeverría, Bolívar. (2010) 2019. *Modernity and "Whiteness."* Reprint, Cambridge: Polity.

Eckstein, Susan E. (1977) 1988. *The Poverty of Revolution: The State and the Urban Poor in Mexico.* Reprint. Princeton, NJ: Princeton University Press.

The Economist. 2012. "The Demographic Transition: More or Less." *The Economist*, September 1, 2012.

El Economista. 2014. "IFE No Castiga al PRI por Tarjetas Soriana." *El Economista*, January 29, 2014.

Edmonds-Poli, Emily, and David A. Shirk. 2009. *Contemporary Mexican Politics*. New York: Rowman and Littlefield.

Ellingwood, Ken. 2012. "Disenchantment May Keep Mexico's Young Voters on Sidelines." *Los Angeles Times*, May 12, 2012.

Elvira Vargas, Rosa. 2004. "Cesa Fox a Ebrard y Figueroa por lo Ocurrido en Ixtayopan." *La Jornada*, December 7, 2004.

———, and Gabriela Romero Sánchez. 2016. "Reforma del DF, Triunfo del Pacto por México, Asegura Peña Nieto." *La Jornada*, January 30, 2016, 28.

Emmerich, Gustavo Ernesto. 2005. *Las Elecciones en la Ciudad de México, 1376–2005*. Mexico City: Instituto Electoral del Distrito Federal and the Universidad Autónoma Metropolitana, 2005.

Enciso L., Angélica. 2011. "Pronasol, 'la Apuesta' que Perdió el Gobierno Federal." *La Jornada*, April 11, 2011.

Engels, Friedrich. (1883) 1940. *The Dialectics of Nature*. Reprint, New York: International Publishers.

Escobar Latapí, Agustín, and Mercedes González de la Rocha. 1995. "Crisis, Restructuring, and Urban Poverty in Mexico." *Environment and Urbanization* 7, no. 1: 57–76.

Esquivel, J. Jesús. 2020. "'Conocíamos las Andanzas de García Luna . . . Pero Debíamos Trabajar con Él': Roberta Jacobson." *Proceso*, May 2, 2020.

Excélsior. 2016. "Clausuran y Decomisan Animales de Parque Cuitláhuac en Iztapalapa." *Excélsior*, October 2, 2016.

Fainstein, Susan. 2010. *The Just City*. Ithaca: Cornell University Press.

Farah, Douglas. 1999. "Prominent Mexican Family Viewed as Threat to U.S." *Washington Post*, June 2, 1999, A1.

Fernandes, Edésio. 2007. "Constructing the 'Right to the City' in Brazil." *Social and Legal Studies* 16, no. 2: 201–19.

Fernández Santillán, José F., Fernando Escalante Gonzalbo, Alicia Ziccardi Contigiani, Pablo Javier Becerra Chávez, and Carlos Sirvent Gutiérrez. 2001. *Análisis y Perspectivas de la Reforma Política del Distrito Federal*. Distrito Federal: Instituto Electoral del Distrito Federal, 2001.

Fernández-Vega, Carlos. 2018. "México SA." *La Jornada*, July 31, 2018.

El Financiero. 2016. "Ismael Figueroa Flores, el Único Independiente que Llega a la Constituyente." *El Financiero*, June 6, 2016.

Fineman, Mark. 1996. "Mexico Corpse Fails to Yield Clues." *Los Angeles Times*, December 20, 1996.

Fischer, Brodwyn, Bryan McCann, and Javier Auyero, eds. 2014. *Cities from Scratch: Poverty and Informality in Urban Latin America*. Durham, NC: Duke University Press.

Fisher, Robert, Yuseph Katiya, Christopher Reid, and Eric Shragge. 2013. "'We are Radical': The Right to the City Alliance and the Future of Community Organizing." *Journal of Sociology and Social Welfare* 40, no. 1: 157–82.

Flores, Mariana. 2017. "The First Constitution of Mexico City." *Metropolis*, February 2, 2017.

Ford, Larry. 1996. "A New and Improved Model of Latin American City Structure." *Geographical Review* 86, no. 3: 437–40.

Forero, Juan. 2006. "Details of Mexico's Dirty Wars from 1960s to 1980s Released." *Washington Post*, November 22, 2006.

Foweraker, Joe, and Todd Landman. 1995. "The August 1994 Elections in Mexico." *Electoral Studies* 14, no. 2: 199–203.

Franco, Diego, Martha Schteingart, and Vicente Ugalde. 2015. "Reseña de la Jornada de Discusión sobre la Reforma Política del Distrito Federal." *Estudios Demográficos y Urbanos* 30, no. 2: 469–92.

Fraser, Nancy. 1997. *Justice Interruptus: Critical Reflections on the "Postsocialist" Condition*. London: Routledge.

Fuentes, David. 2020. "Van en CDMX tras la Ruta del Dinero de los Cárteles." *El Universal*, May 8, 2020.

Gallo, Rubén, ed. 2004. *The Mexico City Reader*. Madison: The University of Wisconsin Press.

Gandler, Stefan. (2015) 2016. *Critical Marxism in Mexico: Adolfo Sánchez Vázquez and Bolívar Echeverría*. Reprint, Chicago: Haymarket Books.

Garcia, Angela. 2015. "Serenity: Violence, Inequality, and Recovery on the Edge of Mexico City." *Medical Anthropology Quarterly* 29, no. 4: 455–72.

García Ochoa, Juan José. 2014. "El Derecho a la Ciudad en el Gobierno del Distrito Federal, Avances y Desafíos." *Dfensor* 12, no. 10: 11–15.

Garza Merodio, Gustavo G. 2006. "Technological Innovation and the Expansion of Mexico City, 1870-1920." *Journal of Latin American Geography* 5, no. 2: 109–26.

Gerlofs, Ben. 2019. "Policing Perception: Postpolitics and the Elusive Everyday." *Urban Geography* 40, no. 3: 378–86.

———. 2021. "Seismic Shifts: Re-centering Geology and Politics in the Anthropocene." *Annals of the American Association of Geographers* 111, no. 3: 828–36.

———. 2022. "Deadly Serious: Humor and the Politics of Aesthetic Transgression." *Dialogues in Human Geography* 12, no. 2: 232–51.

Ghertner, David Asher. 2015. "Why Gentrification Theory Fails in 'Much of the World.'" *City* 19, no. 4: 552–63.

Gilbert, Alan. (1994) 1998. *The Latin American City*. Reprint, London: The Latin American Bureau.

———, and Peter Ward. 1982. "Low-Income Housing and the State." In *Urbanization in Contemporary Latin America: Critical Approaches to the Analysis of Urban Issues*, edited by Alan Gilbert, Jorge E. Hardoy, and Ronaldo Ramírez, 79–127. New York: John Wiley and Sons.

———, and Peter Ward. 1984. "Community Action by the Urban Poor: Democratic Involvement, Community Self-Help or a Means of Social Control?" *World Development* 12, no. 8: 769–82.

Gillingham, Paul, and Benjamin Smith. 2014. "Introduction: The Paradoxes of Revolution." In *Dictablanda: Politics, Work, and Culture in Mexico, 1938–1968*, edited by Paul Gillingham and Benjamin Smith, 1–44. Durham, NC: Duke University Press.

Goldstein, Daniel. 2004. *The Spectacular City: Violence and Performance in Urban Bolivia*. Durham, NC: Duke University Press.

Gómez Ayala, Alexis. 2011. "Huertos Urbanos Contra la Crisis en Ciudad de México." *BBC Mundo*, October 7, 2011.

Gómes Flores, Laura. 2012. "Comienza a Operar el Primer Tramo de la Supervía Poniente." *La Jornada*, October 5, 2012, 41.

González Placencia, Luis. "Message from the President of the CDHDF, Dr. Luis González Placencia, in the Presentation of Recommendation 1/2011." Mexico City: Comisión de Derechos Humanos del Distrito Federal. http://cdhdf. org.mx/2011/01/mensaje-del-presidente-de-la-cdhdf-doctor-luis-gonzalez-placencia-en-la-presentacion-de-la-recomendacion-12011-3/.

Gorz, André. 1967. *Strategy for Labor: A Radical Proposal*. Boston: Beacon Press.

Gray, Neil. 2018. "Beyond the Right to the City: Territorial Autogestion and the Take Over the City Movement in 1970s Italy." *Antipode* 50, no. 2: 319–22.

Gray, Neil, and Libby Porter. 2018. "The Right to the City and Its Limits: Contested Property Claims, Urban Exceptionality, and the Fight for Relational Space in Glasgow's Commonwealth Games 2014." In *Contested Property Claims: What Disagreement Tells Us about Ownership*, edited by Maja Hojer Bruun, Patrick Joseph Cockburn, Bjarke Skærlund Risager, and Mikkel Thorup, 23–38. London: Routledge.

Grayson, George W. 2007. *Mexican Messiah: Andrés Manuel López Obrador*. University Park: Penn State University Press.

GreenTV Noticias. "Para Brugada el Gasto de 114 Mdp en el Parque Cuitláhuac Es 'Muy Poco.'" *GreenTV Noticias*, YouTube, August 12, 2012, 2:00. https://www. youtube.com/watch?v=UBVg2A7sDjo.

Griffin, Ernst, and Larry Ford. 1980. "A Model of Latin American City Structure." *Geographical Review* 70, no. 4: 397–422.

Gugelberger, Georg. 2005. "Waiting for AMLO." *Latin American Perspectives* 32, no. 4: 106–10.

Gunson, Phil. 1999. "Obituary: Mario Ruiz Massieu." *Guardian*, September 19, 1999.

Gutmann, Matthew. 2002. *The Romance of Democracy: Compliant Defiance in Contemporary Mexico*. Berkeley: University of California Press.

Haber, Stephen, Herbert S. Klein, Noel Maurer, and Kevin J. Middlebrook. 2008. *Mexico since 1980*. Cambridge: Cambridge University Press.

Habermas, Jürgen. 1981. "New Social Movements." *Telos*, 49: 33–37.

Hannah, Matthew. 2016. "State Knowledge and Recurring Patterns of State Phobia: From Fascism to Post-Politics." *Progress in Human Geography* 40, no. 4: 476–94.

Hanson, Roger. 1971. *The Politics of Mexican Development*. Baltimore: Johns Hopkins University Press.

Harbers, Imke. 2007. "Democratic Deepening in Third Wave Democracies: Experiments with Participation in Mexico City." *Political Studies* 55, no. 1: 38–58.

Harth Deneke, Jorge Alberto. 1966. "The Colonias Proletarias of Mexico City: Low Income Settlements at the Urban Fringe." PhD diss., MIT.

Harvey, David. 1973. *Social Justice and the City*. Baltimore: Johns Hopkins University Press.

———. 1974. "Population, Resources, and the Ideology of Science." *Economic Geography* 50, no. 3: 256–77.

———. (1984) 2006. *The Limits to Capital*. Reprint, New York: Verso.

———. 2008. "The Right to the City." *New Left Review* 53 (September–October): 23–40.

———. 2012. *Rebel Cities: From the Right to the City to the Urban Revolution*. Brooklyn: Verso.

———. 2014. *Seventeen Contradictions and the End of Capitalism*. Oxford: Oxford University Press.

Hearne, Rory. 2014. "Achieving the Right the City in Practice: Reflections on Community Struggles in Dublin." *Human Geography* 7, no. 3: 14–25.

Henderson, George. 2013. *Value in Marx: The Persistence of Value in a More-than-Capitalist World*. Minneapolis: University of Minnesota Press.

Hernández Chávez, Alicia. 2006. *Mexico: A Brief History*. Berkeley: University of California Press.

Hernandez, Daniel. 2011. *Down and Delirious in Mexico City: The Aztec Metropolis in the Twenty-First Century*. New York: Scribner.

———. 2012. "Calderon's War on Drug Cartels: A Legacy of Blood and Tragedy." *Los Angeles Times*, December 1, 2012.

Hernández León, Simón Alelandro. 2010. "Supervía Poniente: La Democracia en la Ciudad Puesta a Prueba." *La Jornada*, November 20, 2010.

Herrera, Claudia, and Alonso Urrutia A. 2012. "Peña Nieto, PRD, PAN y PRI firman el Pacto por México." *La Jornada*, December 3, 2012.

Hiriart, Pablo. 2019. "La Conspiración de los Críticos: La 4T Se Cura en Salud." *En Financiero*, July 18, 2019.

Hodgson, Dorothy. 2011. "'These Are Not Our Priorities': Maasai Women, Human Rights and the Problem of Culture." In *Gender and Culture at the Limits of Rights*, edited by Dorothy Hodgson, 138–57. Philadelphia: University of Pennsylvania Press.

Holston, James. 2008. *Insurgent Citizenship: Disjunctions of Democracy and Modernity in Brazil*. Princeton, NJ: Princeton University Press.

————. 2009. "Dangerous Spaces of Citizenship: Gang Talk, Rights Talk, and Rule of Law in Brazil." *Planning Theory* 8, no. 1: 12–31.

Holzner, Claudio A. 2007. "The Poverty of Democracy: Neoliberal Reforms and Political Participation of the Poor in Mexico." *Latin American Politics and Society* 49, no. 2: 87–122.

Horowitz, Irving Louis. 1969. "The Norm of Illegitimacy." In *Latin American Radicalism: A Documentary Report on Left and Nationalist Movements*, edited by Irving Louis Horowitz, Josué de Castro, and John Gerassi, 3–29. New York: Vintage Books.

Iaconangelo, David. 2013. "Mario Vargas Llosa, Tells CNN Mexico has Shifted from Dictatorship to Democracy." *Latin Times*, December 23, 2013.

Ibargüengoitia, Jorge. 2004. "Call the Doctor." In *The Mexico City Reader*, edited by Rubén Gallo, 195–97. Madison: University of Wisconsin Press.

Ignatieff, Michael. 2001. *Human Rights as Politics and Idolatry*. Princeton, NJ: Princeton University Press, 2001.

Inclán, María. 2018. *The Zapatista Movement and Mexico's Democratic Transition: Mobilization, Success, and Survival*. New York: Oxford University Press.

Instituto Electoral de la Ciudad de México (IECM). "Resultados del Proceso Electoral Local Ordinario 2017–2018: Alcaldías." Mexico City: Instituto Electoral de la Ciudad de México. http://www.portal.iecm.mx/estadisticas2018/consultas/resultados.php?mod=3.

————. "Ganadores Alcaldías Elección 2021." Mexico City: Instituto Electoral de la Ciudad de México. http://www.iecm.mx/www/Elecciones2021/site/index.html.

Instituto Electoral del Distrito Federal (IEDF). "Resultados de la Elección Local 2006 de Jefe de Gobierno." Mexico City: Instituto Electoral del Distrito Federal. http://www.iedf.org.mx/secciones/elecciones/estadisticas/2006/TOTALES.html?votacion=0.

————. "Estadística de los Resultados 2012." Mexico City: Instituto Electoral del Distrito Federal. http://secure.iedf.org.mx/resultados2012/inicio.php.

————. "Ley de Participación Ciudadana del Distrito Federal." Mexico City: Instituto Electoral del Distrito Federal. http://www.iedf.org.mx/sites/tenemoslaformula/documentos/descarga/ley_de_participacion_ciudadana_2013.pdf.

————. "Elección de la Asamblea Constituyente, Ciudad de México." Mexico City: Instituto Electoral del Distrito Federal. http://computos2016-cdmex.ine.mx/Asambleistas/Entidad/Votos/.

Instituto Nacional de Estadística y Geografía (INEGI). 2017. "México en Cifras." 2017. Mexico City: Instituto Nacional de Estadística y Geografía. http://www.beta.inegi.org.mx/app/areasgeograficas/?ag=15.

Jackson, J. B., Peirce Lewis, David Lowenthal, D. W. Meinig, Marwyn Samuels, David Sopher, and Yi-Fu Tuan. 1979. "Introduction." In *The Interpretation of Ordinary Landscapes: Geographical Essays*, edited by D. W. Meinig, 1–7. Oxford: Oxford University Press.

Jameson, Fredric. 2009. *Valences of the Dialectic*. Brooklyn: Verso.

Jimenez, Edith. 1988. "New Forms of Community Participation in Mexico City: Success or Failure?" *Bulletin of Latin American Research* 7, no. 1: 17–31.

Jiménez, Néstor, and Roberto Garduño. 2018. "Diputados de Morena y PT Defienden Consulta sobre NAICM." *La Jornada*, October 25, 2018.

Johns, Michael. 1997. *The City of Mexico in the Age of Díaz*. Austin: University of Texas Press.

La Jornada. 1997. "Arrasó." *La Jornada*, July 7, 1997.

———. 2000. "Refrenda la Ciudad su Confianza al PRD." *La Jornada*, July 3, 2000.

———.2009. "La Jornada 88." *La Jornada*, May 4, 2009.

———. 2018."Banco Suizo Teme que AMLO Utilice el Referendo para Ampliar Su Sexenio." *La Jornada*, October 30, 2018, 6.

———. 2021. "Desafuero: Aniversario de una Infamia." *La Jornada*, April 9, 2021.

Joseph, Gilbert, and Jürgen Buchenau. 2013. *Mexico's Once and Future Revolution: Social Upheaval and the Challenge of Rule since the Late Nineteenth Century*. Durham, NC: Duke University Press, 2013.

Kandell, Jason. 1988. *La Capital: The Biography of Mexico City*. New York: Random House.

———. (1988) 1996. "Mexico's Megalopolis." In *I Saw a City Invincible: Urban Portraits of Latin America*, edited by Gilbert M. Joseph and Mark D. Szuchman, 181–201. Reprint, Wilmington: Scholarly Resources Inc.

Kanai, J. Miguel, and Iliana Ortega-Alcázar. 2009. "The Prospects for Progressive Culture-Led Urban Regeneration in Latin America: Cases from Mexico City and Buenos Aires." *International Journal of Urban and Regional Research* 33, no. 2: 483–501.

Kimmelman, Michael. 2017. "Mexico City, Parched and Sinking, Faces a Water Crisis." *New York Times*, February 17, 2017.

Kipfer, Stefan, Parastou Saberi, and Thorben Wieditz. 2013. "Henri Lefebvre: Debates and Controversies." *Progress in Human Geography* 37, no. 1: 115–134.

Kitroeff, Natalie, Maria Abi-Habib, James Glanz, Oscar Lopez, Weiyi Cai, Evan Grothjan, Miles Peyton, and Alejandro Cegarra. 2021. "Why the Mexico City Metro Collapsed." *New York Times*, June 12, 2021.

Knight, Alan. 1986. *The Mexican Revolution, Volume II: Counter-Revolution and Reconstruction*. Cambridge: Cambridge University Press.

———. 1990. "Historical Continuities in Social Movements." In *Popular Movements and Political Change in Mexico*, edited by Joe Foweraker and Ann L. Craig, 78–102. Boulder: Lynn Rienner Publishers.

———. 2012. "Mexico: Democracy Interrupted by Jo Tuckman—Review." *Guardian*, June 29, 2012.

———, and Wil Pansters, eds. 2006. *Caciquismo in Twentieth-Century Mexico*. London: Institute for the Study of the Americas.

Kofman, Eleonore, and Elizabeth Lebas. 1996. "Lost in Transposition: Time, Space, and the City." In *Writings on Cities*, edited and translated by Eleanore Kofman and Elizabeth Lebas, 3–62. Malden: Blackwell.

Kohout, Michal. 2009. "Hegemonic Geographies of the Mexican Neoliberal State." *Human Geography* 2, no. 1: 45–62.

Laclau, Ernesto, and Chantal Mouffe. 2001. *Hegemony and Socialist Strategy*. New York: Verso.

Langner, Ana. 2017. "En Debate Público, la Constitución CDMX." *El Economista*, June 12, 2017.

Langston, Joy. 2017. *Democratization and Authoritarian Party Survival: Mexico's PRI*. New York: Oxford University Press.

Latour, Bruno. 1987. *Science in Action: How to Follow Scientists and Engineers through Society*. Milton Keynes: Open University Press.

La Trobe, Charles J. 1834. *The Rambler in Mexico*. London: R. B. Seeley and W. Burnside.

Leal Martínez, Alejandra. 2020. "Securing the Street: Urban Renewal and the Fight against "Informality" in Mexico City." In *Futureproof: Security Aesthetics and the Management of Life*, edited by David Asher Ghertner, Hudson McFann, and Daniel Goldstein, 245–270. Durham, NC: Duke University Press.

Lear, John. 2001. *Workers, Neighbors, and Citizens: The Revolution in Mexico City*. Lincoln: University of Nebraska Press.

Lefebvre, Henri. (1974) 1991. *The Production of Space*. Reprint, Malden: Blackwell.

———. (1968) 1996. "The Right to the City." In *Writings on Cities*, edited and translated by Eleanore Kofman and Elizabeth Lebas, 61–181. Reprint, Malden: Blackwell.

———. (1970) 2003(a). *The Urban Revolution*. Reprint, Minneapolis: University of Minnesota Press.

———. (1986) 2003(b). "Becoming and the Historical." In *Henri Lefebvre: Key Writings*, edited by Stuart Elden, Elizabeth Lebas, and Eleonore Koman, 65–66. Reprint, New York: Continuum.

Leñero, Vicente. 2004. "La Diana." In *The Mexico City Reader*, edited by Rubén Gallo, 152–162. Madison: University of Wisconsin Press.

Leontidou, Lila. 2010. "Urban Social Movements in 'Weak' Civil Societies: The Right to the City and Cosmopolitan Activism in Southern Europe." *Urban Studies* 47, no. 6: 1179–203.

Lewis, Oscar. 1959. *Five Families: Mexican Case Studies in the Culture of Poverty*. New York: Basic Books.

———. 1961. *The Children of Sánchez: Autobiography of a Mexican Family*. New York: Random House.

Lida, David. 2008. *First Stop in the New World: Mexico City, Capital of the Twenty-First Century*. New York: Riverhead Books.

Llanos Samaniego, Raúl. 2016. "Presidirá Alejandro Encinas la Constituyente." *La Jornada*, 5 October 5, 2016.

———, Alejandro Cruz Flores, Laura Gómez Flores, and Bertha Teresa Ramírez. 2015. "La Gente Dijo No al Corredor." *La Jornada*, December 7, 2015.

Lombard, Melanie. 2013. "Citizen Participation in Urban Governance in the Context of Democratization: Evidence from Low-Income Neighbourhoods in Mexico." *International Journal of Urban and Regional Research* 37, no. 1: 135–50.

———. 2014. "Constructing Ordinary Places: Place-Making in Urban Informal Settlements in Mexico." *Progress in Planning* 94: 1–53.

Lomnitz, Claudio. 2003. "Times of Crisis: Historicity, Sacrifice, and the Spectacle of Debacle in Mexico City." *Public Culture* 15, no. 1: 127–47.

———. 2005. *Death and the Idea of Mexico*. Brooklyn, Zone Books.

Lopez, Rene, and Jaime Aviles. 2005. ""Cobarde," el Proceder de PGR y Dos Panistas." *La Jornada*, April 20, 2005.

López Obrador, Andrés Manuel. 2006. "Aquí está la Muestra de lo que Somos y de lo que Seremos Capaces de Llevar a Cabo: AMLO." *La Jornada*, November 21, 2006.

López Orozco, Leticia. 2014. "The Revolution, Vanguard Artists and Mural Painting." *Third Text* 28, no. 3: 256–68.

López-Morales, Ernesto, Hyun Bang Shin, and Loretta Lees. 2016. "Latin American Gentrifications." *Urban Geography* 37, no. 8: 1091–1108.

MacLeod, Gordon. 2011. "Urban Politics Reconsidered: Growth Machine to Post-Democratic City?" *Urban Studies* 48, no. 12: 2629–60.

Mancebo, François. 2007. "Natural Hazards and Urban Policy in Mexico City." *Journal of Alpine Research* 95, no. 2: 108–18.

Marcuse, Peter. 2009. "From Critical Urban Theory to the Right to the City." *City* 13, no. 2–3: 185–97.

Martínez, Fabiola. 2015. "Canceló Gobernación el Acceso Directo a los Archivos sobre al Guerra Sucia." *La Jornada*, March 11, 2015.

———, and Matilde Pérez. 2000. "Renuncia Formal de Porfirio Muñoz Ledo al PRD." *La Jornada*, January 13, 2000.

Martínez, Oscar J. 2016. *Mexico's Uneven Development: The Geographical and Historical Context of Inequality*. New York: Routledge.

Martinez, Ricardo, Tim Bunnell, and Michele Acuto. 2020. "Productive Tensions? The 'City' Across Geographies of Planetary Urbanization and the Urban Age." *Urban Geography* 42, no. 7: 1011–22.

Marx, Karl. (1867) 1967(a). *Capital: A Critique of Political Economy*, vol. I. Reprint, New York: International Publishers.

———. (1873) 1967(b). "Afterword to the Second German Edition." In *Capital: A Critique of Political Economy*, vol. I, by Karl Marx, 22–29. Reprint, New York: International Publishers.

———. (1859) 1970. *A Contribution to the Critique of Political Economy*. Reprint, New York: International Publishers.

Massey, Doreen. 2005. *For Space*. London: Sage.

De Mauleón, Héctor. 2015. "La Ciudad de los Palacios." *Nexos*, April 24, 2015.

McAdams, A. James. 2017. *Vanguard of the Revolution: The Global Idea of the Communist Party*. Princeton, NJ: Princeton University Press.

McCarthy, James. 2002. "First World Political Ecology: Lessons from the Wise Use Movement." *Environment and Planning A*, 34, no. 7: 1281–1302.

———. 2013. "We Have Never Been "Post-Political."" *Capitalism, Nature, Socialism* 24, no. 1: 19–25.

McCormack, Derek. 2012. "Geography and Abstraction: Towards an Affirmative Critique." *Progress in Human Geography* 36, no. 6: 715–34.

McFarlane, Colin, and Jonathan Silver. 2017. "Navigating the City: Dialectics of Everyday Urbanism." *Transactions of the Institute of British Geographers* 43, no. 3: 458–71.

Medina Ramírez, Salvador. 2015. "La ZODE Chapultepec: ¿Operación Inmobiliaria, Espacio Público o Centro Comercial Privado?." *Horizontal*, August 5, 2015.

Melchor, Fernando. "The Political Reform of Mexico City and the 'Crowdsourcing' of a New Constitution." *Obsidian Mirror*, September 23, 2016. https://obsidian-mirror.net/blog/2017/1/27/the-political-reform-of-mexico-city-and-the-crowdsourcing-of-a-new-constitution.

Melimopoulos, Elizabeth. "Mexico's National Guard: What, Who and When." *Aljazeera*, June 30, 2019. Aljazeera.com/news/2019/6/30/mexicos-national-guard-what-who-and-when.

Méndez, Ernesto. 2016. "Clara Brugada Abrió Granja de Crueldad Animal." *Excélsior*, June 23, 2016.

Méndez, Enrique. 2018. "Obedeceré el Mandato Ciudadano: AMLO." *La Jornada*, October 30, 2018, 2.

Merrifield, Andy. 2006. *Henri Lefebvre: A Critical Introduction*. New York: Routledge.

———. 2011. "The Right to the City and Beyond." *City* 15, no. 3–4: 468–76.

———. 2014. *The New Urban Question*. London, Pluto Press.

Mier M. M., T. Rocha, and C. A. Rabell Romero. 1988. "Ciudad de México: Características socioeconómicas de los damnificados de los sismos de septiembre." In *Atlas de la Ciudad de México*, vol. 5, edited by G. Garza, 162–66. Mexico City: Departamento del Distrito Federal and El Colegio de México.

Mitchell, Don. 1996. *The Lie of the Land: Migrant Workers and the California Landscape*. Minneapolis: University of Minnesota Press.

———. 2003. *The Right to the City: Social Justice and the Fight for Public Space*. New York: Guilford Press.

———. 2008. "New Axioms for Reading the Landscape: Paying Attention to Political Economy and Social Justice." In *Political Economies of Landscape Change*, edited by James Wescoat and Douglas Johnston, 29–50. Dordrecht: Springer.

———. "Tent City: Lessons on the Right to the City from the Urban Interstices."
SKORFoundation, YouTube, December 6, 2011, 1:03:13. https://www.youtube.
com/watch?v=czbJJ4iphVQ.

Mitchell, Don, and Nik Heynan. 2009. "The Geography of Survival and the
Right to the City: Speculations on Surveillance, Legal Innovation, and the
Criminalization of Intervention." *Urban Geography* 30, no. 6: 611–32.

Mitchell, Don, Lynn Staeheli, and Kafui Attoh. 2015. "Whose City? What Politics?
Contentious and Non-Contentious Spaces on Colorado's Front Range." *Urban
Studies* 52, no. 14: 2633–48.

Moctezuma, Pedro. 2007. "Community-Based Organization and Participatory
Planning in South-East Mexico City." *Environment and Urbanization* 13, no. 2:
117–33.

Moguel, Julio. 2007. "Salinas' Failed War on Poverty." *NACLA*, September 25,
2007.

Monsiváis, Carlos. 1987. *Entrada Libre: Crónicas de la Sociedad que se Organiza.*
Mexico City: Ediciones Era.

———. 1997. *Mexican Postcards.* New York: Verso.

Mora, Karla. 2012. "Reviven Parque Cuitláhuac en lo que Antes Era Basurero." *El
Universal*, August 13, 2012.

Moreno, Louis. 2014. "The Urban Process Under Financialized Capitalism." *City*
18, no. 3: 244–68.

Moreno-Brid, Juan Carlos, and Jaime Ros. 2009. *Development and Growth in
the Mexican Economy: A Historical Perspective.* New York: Oxford University
Press.

Morse, Julie. 2015. "Why Residents of Mexico City are Organizing Themselves
Against Parking Meters." *Vice*, August 14, 2015.

Mouffe, Chantal. 2005. *On the Political.* New York: Routledge.

Moyn, Samuel. 2010. *The Last Utopia: Human Rights in History.* Cambridge: The
Belknap Press of Harvard University.

Moynihan, Colin. 2012. "Owner Had Right to Clear Zucotti Park, Judge Says."
New York Times, April 8, 2012.

Müller, Markus-Michael. 2013. "Penal Statecraft in the Latin American City:
Assessing Mexico City's Punitive Urban Democracy." *Social and Legal Studies*
22, no. 4: 441–63.

Muñoz, Alma. 2015(a). "El Senador Alejandro Encinas hizo Efectiva su Renuncia
al PRD." *La Jornada*, January 23, 2015(a).

———. 2015(b). "Renuncia Ebrard al PRD, Reprocha Cercanía del Partido con
Peña." *La Jornada*, February 27, 2015(b).

———, and Georgina Saldierna. 2014. "Cárdenas deja el PRD, Partido que Fundó
Hace 25 Años." *La Jornada*, November 25, 2014.

Murray, Martin. 2017. *The Urbanism of Exception: The Dynamics of Global City
Building in the Twenty-First Century.* Cambridge: Cambridge University Press.

Nelson, Lise. 2003. "Decentering the Movement: Collective Action, Place, and the

'Sedimentation' of Radical Political Discourses." *Environment and Planning D: Society and Space* 21, no. 5: 559–81.

Neria, Leticia, and Mark Aspinwall. 2015. "Popular Comics and Authoritarian Injustice Frames in Mexico." *Latin American Research Review* 51, no. 1: 22–42.

Neufeld, Stephen. 2012. "Behaving Badly in Mexico City: Discipline and Identity in the Presidential Guards, 1900–1911." In *Forced Marches: Soldiers and Military Caciques in Modern Mexico*, edited by Ben Fallow and Terry Rugeley, 81–109. Tucson: University of Arizona Press.

New York Times. 1910. "Porfirio Diaz of Mexico: The Life and Work of the Master-Builder of a Great Commonwealth Set Forth in an Entertaining New Volume by Jose F. Godoy." *New York Times*, March 19, 1910, BR1.

———. 1911. "The Attitude of Diaz." *New York Times*, May 10, 1911, 10.

———. 2000. "Mexico's Democratic Breakthrough." *New York Times*, July 4, 2000.

———. 2001. "Carlos Hank González, 73, Veteran Mexican Politician." *New York Times*, August 13, 2001.

Nilsen, Alf Gunvald. 2009. "'The Authors and the Actors of their Own Drama': Towards a Marxist Theory of Social Movements." *Capital and Class* 33, no. 3: 109–39.

Norton, Richard. 2003. "Feral Cities." *Naval War College Review* 56, no. 4: 97–106.

Nuñez, Ernesto. 2017. "'Hay Ignorancia y Mala Fe.'" *Reforma*, March 19, 2017.

Olayo, Ricardo, and Raúl Llanos. 1999. "Cárdenas Renunciará como Jefe de Gobierno." *La Jornada*, September 28, 1999.

Ollman, Bertell. 1971. *Alienation: Marx's Conception of Man in Capitalist Society*. New York: Cambridge University Press, 1971.

———. 1993. *Dialectical Investigations*. New York: Routledge.

———. 2003. *Dance of the Dialectic: Steps in Marx's Method*. Urbana: University of Illinois Press.

Ong, Aihwa. 2011. "Worlding Cities, or the Art of Being Global." In *Worlding Cities: Asian Experiments and the Art of Being Global*, edited by Ananya Roy and Aihwa Ong, 1–26. Oxford: Wiley Blackwell.

Ortiz Flores, Enrique. 1990. *Annual Report*. Ottawa: Habitat International Coalition, May 1990.

———. 1995. "Carta por los Derechos a la Ciudad y la Vivienda." In *Homenaje a Enrique Ortíz*, edited by Casa y Ciudad, 32–43. Mexico City: Casa y Ciudad, 1995.

———. 2008. *The Right to the City as Complex System: Repercussions for the Formulation of the Charter*. Mexico City: Habitat International Coalition-Latin America, June 2008.

O'Toole, Gavin. 2003. "A New Nationalism for a New Era: The Political Ideology of Mexican Neoliberalism." *Bulletin of Latin American Research* 22, no. 3: 269–90.

Páramo, Arturo. 2012. "Presenta Iztapalapa Parque Cuitláhuac a Alcaldes de Latinoamérica." *Excélsior*, September 11, 2012.

Park, Robert. 1928. "Human Migration and the Marginal Man." *American Journal of Sociology* 33, no. 6: 881–93.

———, and Ernest Burgess. 1925. *The City: Suggestions for Investigation of Human Behavior in the Urban Environment*. Chicago: University of Chicago Press.

Parnell, Susan, and Edgar Pieterse. 2010. "The 'Right to the City': Institutional Imperatives of a Developmental State." *International Journal of Urban and Regional Research* 34, no. 1: 146–62.

Parnreiter, Christof. 2015. "Strategic Planning, the Real Estate Economy, and the Production of New Spaces of Centrality. The Case of Mexico City." *Erdkunde: Archive for Scientific Geography* 69, no. 1: 21–31.

Paz, Octavio. 1985. *The Labyrinth of Solitude and The Other Mexico; Return to the Labyrinth of Solitude; Mexico and the United States; The Philanthropic Ogre*. New York: Grove Weidenfeld.

Patomäki, Heikki. 2017. "Praxis, Politics, and the Future: A Dialectical Critical Realist Account of World-Historical Causation." *Journal of International Relations and Development* 20, no. 4: 805–25.

Pensado, Jaime. 2013. *Rebel Mexico: Student Unrest and Authoritarian Political Culture During the Long Sixties*. Redwood City, CA: Stanford University Press.

Perez-Pena, Richard. 1994. "The Assassination in Mexico; Luis Colosio: A Party Man for 2 Decades." *New York Times*, March 25, 1994.

Perlman, Janice. 1975. "The Myth of Marginality." *Politics and Society* 5, no. 2: 131–60.

Perló Cohen, Manuel. 1979. "Política y Vivienda en México 1910–1952." *Revista Mexicana de Sociología* 41, no. 3: 769–835.

———. 1993. "La Ciudad de México: Ese Pretendido Problema." *Ibero-amerikanisches Archiv* 19, no. 1/2: 123–150.

Piccato, Pablo. 2001. *City of Suspects: Crime in Mexico City, 1900–1931*. Durham, NC: Duke University Press.

Pichardo, Nelson A. 1997. "New Social Movements: A Critical Review." *Annual Review of Sociology* 23, no. 1: 411–30.

Pickvance, Chris. 2003. "From Urban Social Movements to Urban Movements: A Review and Introduction to a Symposium on Urban Movements." *International Journal of Urban and Regional Research* 27, no. 1: 102–9.

Pierce, Joseph, Olivia R. Williams, and Deborah G. Martin. 2016. "Rights in Places: An Analytical Extension of the Right to the City." *Geoforum* 70: 79–88.

Poniatowska, Elena. (1988) 1995. *Nothing, Nobody: The Voices of the Mexico City Earthquake*. Reprint, Philadelphia, PA: Temple University Press.

———. 2000. "Forward." In *Mexico City: A Cultural and Literary Companion*, edited by Nick Caistor, vii–ix. Oxford: Signal Books.

———. (1971) 2012. *La Noche de Tlatelolco*. Reprint, Mexico City: Ediciones Era.

Portes, Alejandro, and John Walton. 1976. *Urban Latin America: The Political Condition from Above and Below*. Austin: University of Texas Press.

Proceso. 2005. "Histórica "Marcha del Silencio": López Obrador Propone un Pacto Social Incluyente." *Proceso*, April 25, 2005.

Progressive Alliance. "Defend Mexico City's First Political Constitution." Berlin: Progressive Alliance. http://progressive-alliance.info/2017/03/15/defend-mexico-citys-first-political-constitution/.

Purcell, Mark. 2002. "Excavating Lefebvre: The Right to the City and Its Urban Politics of the Inhabitant." *Geojournal* 58, no. 2: 99–108.

———. 2013. "To Inhabit Well: Counterhegemonic Movements and the Right to the City." *Urban Geography* 34, no. 4: 560–74.

———. 2014. "Possible Worlds: Henri Lefebvre and the Right to the City." *Journal of Urban Affairs* 36, no. 1: 141–54.

Quintero Morales, Josefina. 2012. "Inauguran en Iztapalapa el Parque Cuitláhuac, el Quinto Más Grande del DF." *La Jornada*, August 13, 2012.

———. 2017. "Sin la Presencia de Grupos Sociales, Gómez Gallardo Informa de Labores." *La Jornada*, April 5, 2017, 31.

Rabell García, Enrique. 2017. "La Reforma Política de la Ciudad de México." *Cuestiones Constitucionales* 36, Enero-Junio: 243–70.

Rakowski, Cathy, ed. 1994. *Contapunto: The Informal Sector Debate in Latin America*. Albany: SUNY Press.

Rama, Anahi, and Lizbeth Diaz. 2014. "Violence Against Women 'Pandemic' in Mexico." *Reuters*, March 7, 2014.

Rancière, Jacques. 1994. "Post-Democracy, Politics and Philosophy: An Interview with Jacques Rancière." *Angelaki*: 171–78.

———, Davide Panagia, and Rachel Bowlby. 2001. "Ten Theses on Politics." *Theory and Event* 5, no. 3. doi: 10.1353.tae.2001.0028.

———. 2007. *On the Shores of Politics*. New York: Verso.

———. 2010. *Dissensus: On Aesthetics and Politics*. New York: Continuum.

Rasmussen, Anthony. 2017. "Sales and Survival within the Contested Acoustic Territories of Mexico City's Historic Centre." *EtÚomusicology Forum* 26, no. 3: 307–30.

Regeneración. 2015. "IEDF, PRD y GDF Coludidos para Cacer Fraude en Consulta por Corredor Chapultepec." *Regeneración*, December 2, 2015.

Rello Gómez, Jaime. "Palabras de Jaime Rello a Nombre del Comité Promotor." *Mexico City: Derecho a la Ciudad DF* (blog). http://derechoalaciudaddf. blogspot.com/2010/07/palabras-de-jaime-rello-nombre-del.html.

The Right to the City Alliance. "Member Organizations." Brooklyn: The Right to the City Alliance. http://righttothecity.org/members.

Roberts, Bryan. 2005. "Globalization and Latin American Cities." *International Journal of Urban and Regional Research* 29, no. 1: 110–23.

Robinson, Jennifer. 2016. "Thinking Cities through Elsewhere: Comparative Tactics for a More Global Urban Studies." *Progress in Human Geography* 40, no. 1: 3–29.

Rodríguez Kuri, Ariel. 2010. *Historia del Desasosiego: La Revolución en la Ciudad de México, 1911–1922*. Mexico City: El Colegio de México.

———. 2012. "Introducción." In *Historia política de la Ciudad de México (desde su Fundación hasta el año 2000)*, edited by Ariel Rodríguez Kuri, 9–18. Mexico City: El Colegio de México.

Romero, Gabriela, and Alejandro Cruz. 2010. "Dan Otro Paso para que el DF Tenga Constitución." *La Jornada*, July 14, 2010, 24.

Rorty, Richard. 1996. "What's Wrong with 'Rights.'" *Harper's Magazine* 292, no. 1753: 15–18.

Rosales, Maricela. 2018. "Conspiraciónes contra AMLO." *Animal Político*, April 9, 2018.

Ross, John. 2008. "The Demise of Mexico's PRD." *Counterpunch*, May 17, 2008.

———. 2009. *El Monstruo: Dread and Redemption in Mexico City*. New York: Nation Books.

Rossell, Daniela. 2004. "Las Lomas II." In *The Mexico City Reader*, edited by Rubén Gallo, 114–20. Madison: University of Wisconsin Press.

Roy, Ananya, and Nezar Alsayyad, eds. 2004. *Urban Informality: Transnational Perspectives from the Middle East, Latin America, and South Asia*. Lanham, MD: Lexington Books.

Ruiz, Patricio. 2015. "La Falacia del Corredor Cultural Chapultepec." *Arquine*, August 4, 2015.

Salgado, Agustin. 2010. "En Favor de la Supervía, 74% de Mil 396 Encuestados, Revela un Sondeo." *La Jornada*, August 12, 2010.

———, and Angel Bolaños Sanchez. 2005. "AMLO: Impediremos toda Chicanada Dilatoria de PGR." *La Jornada*, April 16, 2005.

Sánchez, Carlos Alberto, and Robert Eli Sanchez, eds. 2017. *Mexican Philosophy in the 20th Century: Essential Readings*. Oxford: Oxford University Press.

Sassen, Saskia. (1991) 2001. *Global Cities: New York, London, Tokyo*. Reprint, Princeton, NJ: Princeton University Press.

———. (1994) 2012. "A New Geography of Centers and Margins." In *Cities in a World Economy*, edited by Saskia Sassen, 323–29. Reprint, Thousand Oaks, CA: Sage.

Schmid, Christian. 2012. "Henri Lefebvre, the Right to the City, and the New Metropolitan Mainstream." In *Cities for People, Not for Profit: Critical Urban Theory and the Right to the City*, edited by Neil Brenner, Peter Marcuse, and Margit Mayer, 42–62. New York: Routledge.

Scott, D. C. 1993. "Mexico City Mayor Steers Slow Course to Self-Rule." *Christian Science Monitor*, April 22, 1993.

Scruggs, Gregory. 2017. "The People Power behind Mexico City's New Constitution." *Citiscope*, February 3, 2017.

Sharpe, Christina. 2016. *In the Wake: On Blackness and Being*. Durham, NC: Duke University Press.

Sheppard, Randal. 2011. "Nationalism, Economic Crisis and 'Realistic Revolution' in 1980s Mexico." *Nations and Nationalism* 17, no. 3: 500–19.

Sieff, Kevin, and Shayna Jacobs. 2020. "Mexico Welcomes U.S. Release of Accused Former Defense Minister, but Episode has Deepened Mistrust on both Sides." *Washington Post*, November 19, 2020.

Siembieda, William. 1996. "Looking for a Place to Live: Transforming the Urban Ejido." *Bulletin of Latin American Research* 15, no. 3: 371–85.

Sin Embargo. 2012. "Rosario Robles: Del Éxito Político a Víctima del Amor . . . y de Ahí, al Equipo de Peña Nieto." *Sin Embargo*, September 11, 2012.

Slater, Tom. 2021. *Shaking Up the City: Ignorance, Inequality, and the Urban Question.* Berkeley: The University of California Press.

Smart, Alan, and Josephine Smart. 2017. "Ain't Talkin' 'Bout Gentrification: The Erasure of Alternative Idioms of Displacement Resulting from Anglo-American Academic Hegemony." *International Journal of Urban and Regional Research* 41, no. 3: 518–25.

Smith, Neil. 2003. "Foreword." In *The Urban Revolution*, by Henri Lefebvre, vii–xxiii. Minneapolis: University of Minnesota Press.

Smith, Peter. 1989. "The 1988 Presidential Succession in Historical Perspective." In *Mexico's Alternative Political Futures*, edited by Wayne Cornelius, Judith Gentleman, and Peter Smith, 391–415. San Diego, CA: Center for US-Mexican Studies.

Soares, Sergei, Rafel Guerreiro Osório, Fábio Veras Soares, Marcelo Medeiros, and Eduardo Zepeda. 2007. "Conditional Cash Transfers in Brazil, Chile, and Mexico: Impacts upon Inequality." Working Paper no. 35, International Poverty Centre, April 2007. http//www.ipcig.org/pub/IPCWorkingPaper35.pdf.

Soederberg, Susanne. 2001. "From Neoliberalism to Social Liberalism: Situations the National Solidarity Program within Mexico's Passive Revolutions." *Latin American Perspectives* 28, no. 3: 104–23.

Stolle-McAllister, John. 2005. "What Does Democracy Look Like? Local Movements Challenge the Mexican Transition." *Latin American Perspectives* 32, no. 4: 15–35.

Story, Dale. 1986. *Industry, the State, and Public Policy in Mexico.* Austin: University of Texas Press.

Story, Louise, and Stephanie Saul. 2015. "Stream of Foreign Wealth Flows to Elite New York Real Estate." *New York Times*, February 7, 2015.

Suárez, Gerardo. 2017. "Asamblea Constituyente Aprueba la Primera Constitución de la CDMX." *El Universal*, January 31, 2017.

Los Supercívicos. "El Jefe del Defe." Los Supercívicos, YouTube, December 2, 2015, 5:38.
https://www.youtube.com/watch?v=GX92P2Fv_gk.

Swyngedouw, Erik. 2009. "The Antinomies of the Postpolitical City: In Search of a Democratic Politics of Environmental Production." *International Journal of Urban and Regional Research* 33, no. 3: 601–20.

———. 2011. "Interrogating Post-Democratization: Reclaiming Egalitarian Political Spaces." *Political Geography* 30, no. 7: 370–80.

———. "Erik Swyngedouw in Conversation with Neil Smith." Center for Place Culture Politics, Vimeo, April 18, 2012, 1:37:15. https://vimeo.com/42628112.

Tamayo, Sergio. 2006. "Espacios de Ciudadanía, Espacios de Conflicto." *Sociológica* 21, no. 61: 11–40.

———. 2015. "La Participación Ciudadana: Un Proceso." *Revista Mexicana de Opinión Pública* 18, (January–June): 157–83.

Tenorio-Trillo, Mauricio. 1996. *Mexico at the World's Fairs: Crafting a Modern Nation.* Berkeley: University of California Press.

———. 2012. *I Speak of the City: Mexico City at the Turn of the Twentieth Century.* Chicago: University of Chicago Press.

———. 2020. "History on Foot: Walking Mexico City." *Current History* 119, no. 814: 66–72.

Thompson, Warren. 1929. "Population." *American Journal of Sociology* 34, no. 6: 959–75.

Thornton, Christy. 2021. *Revolution in Development: Mexico and the Governance of the Global Economy.* Oakland: University of California Press.

Torres, Rubén. 2015. "Morena Acusa Compra de Votos a Favor del Corredor Chapultepec." *El Economista*, December 2, 2015.

Tribunal Electoral del Poder Judicial de la Federación. "Ley de Participación Ciudadana del Distrito Federal (Ciudad de México)." Mexico City: Tribunal Electoral del Poder Judicial de la Federación. http://www.trife.gob.mx/legislacion-jurisprudencia/catalogo/2015-ley-de-participacion-ciudadana-del-distrito-f.

Tuckman, Jo. 2000. "Mexicans Toast Fox Victory." *Guardian*, July 3, 2000.

———. 2001. "Carlos Hank Gonzales: Powerbroker Behind Mexico's Ruling Elite." *Guardian*, August 14, 2001.

———. 2012(a). *Mexico: Democracy Interrupted.* New Haven, CT: Yale University Press.

———. 2012(b). "Mexico Elections: Claims of Dirty Tricks Cast Shadow over Peña Nieto's Victory." *Guardian*, July 4, 2012.

Tuiran, Rodolfo, Virgilio Partida, Octavio Mojarro, and Elena Zúñiga. "Fertility in Mexico: Trends and Forecast." New York: United Nations. https://www.un.org/development/desa/pd/sites/www.un.org.development.desa.pd/files/unpd_egm_200203_countrypapers_fertility_in_mexico_tuiran_partida_mojarro_zuniga.pdf.

Tushnet, Mark. 1984. "An Essay on Rights." *Texas Law Review* 62, no. 8: 1363–1412.

Uitermark, Justus, Walter Nicholls, and Maarten Loopmans. 2012. "Cities and Social Movements: Theorizing Beyond the Right to the City." *Environment and Planning A* 44, no. 11: 2546–54.

United Nations. *World Urbanization Prospects, 2018 Revision.* New York: United Nations. https://population.un.org/wup/Download/.

———. *National Accounts Main Aggregates Database.* New York: United Nations. https://unstats.un.org/unsd/snaama/dnllist.asp.

———. *World Population Prospects.* New York: United Nations. https://esa.un.org/unpd/wpp/.

United States General Accounting Office (USGOA). 1998. "Raul Salinas, Citibank, and Alleged Money Laundering." *Report to the Ranking Minority Member, Permanent Subcommittee on Investigations, Committee on Governmental Affairs, U.S. Senate.* Report no. B-281327, October 30, Washington, DC.

Van Ramshorst, Jared. 2019. "Laughing about It: Emotional and Affective Spaces of Humour in the Geopolitics of Migration." *Geopolitics* 24, no. 4: 896–915.

Van Sant, Levi, Elizabeth Hennessy, Mona Domosh, Mohammed Rafi Arefin, Nathan McClintock, and Sharlene Mollett. 2020. "Historical Geographies of, and for, the Present." *Progress in Human Geography* 44, no. 1: 168–88.

Varela, Micaela. 2021. "Rosario Robles Retira su Oferta de Declararse Culpable y Decide Ir a Juicio." *El País*, March 10, 2021.

Varley, Ann. 1987. "The Relationship between Tenure Legalization and Housing Improvements: Evidence from Mexico City." *Development and Change* 18: 463–81.

Vasconcelos Calderón, José. (1925) 1997. *La Raza Cósmica (The Cosmic Race): Bilingual Edition.* Reprint, Baltimore: Johns Hopkins University Press.

Vela, David Saúl. 2021. "Rosario Robles Ofrece Declararse Culpable y Recibir Seis Años de Prisión por la Estafa Maestra." *El Financiero*, March 3, 2021.

Velasquez, Tamara. "No, Mexico City Is Not 'the New Berlin.'" *Medium*, April 18, 2017. https://medium.com/@tamaravelasquez/no-mexico-city-is-not-the-new-berlin-a-response-to-vice-ca677296c417.

Villalobos, Horacio. 2002–2013. *Desde Gayola* [TV series]. Mexico City: Telehit (2001–2006); 52MX (2008–2013).

Villavicencio, Diana. 2015. "Critica PRI Corredor Cultural Chapultepec." *El Universal*, November 11, 2015.

Vitz, Matthew. 2018. *A City on a Lake: Urban Political Ecology and the Growth of Mexico City*. Durham, NC: Duke University Press.

Vradis, Antonis. "Intervention—the Right Against the City." *Antipode Online.* https://antipodefoundation.org/2012/10/01/intervention-the-right-against-the-city/.

Wacquant, Loïc. 2008. *Urban Outcasts: A Comparative Sociology of Advanced Marginality*. Malden: Polity Press.

Wainwright, Oliver. 2014. "The Truth about Property Developers: How they are Exploiting Our Planning Authorities and Ruining Our Cities." *Guardian*, September 17, 2014.

Walker, Richard A. 1977. "The Suburban Solution: Urban Geography and Urban Reform in the Capitalist Development of the United States." PhD diss., Johns Hopkins University.

Ward, Peter. 1978. "Self-Help Housing in Mexico City: Social and Economic Determinants of Success." *The Town Planning Review* 49, no. 1: 38–50.

———. 1989. "Government without Democracy in Mexico City: Defending the High Ground." In *Mexico's Alternative Political Futures*, edited by Wayne A.

Cornelius, Judith Gentleman, and Peter H. Smith, 307–23. San Diego: Center for US-Mexican Studies.

———. 1990. *Mexico City: The Production and Reproduction of an Urban Environment*. Boston: G. K. Hall.

———. 2004. *México Megaciudad: Desarrollo y Política 1970–2002*. Mexico City: Miguel Angel Porrúa.

Weiss, Sandra. "Mexico City's 'Crowdsourced' Constitution." *International Politics and Society*, April 27, 2017.

Wigle, Jill, and Lorena Zárate. 2010. "Mexico City Creates Charter for the Right to the City." *Progressive Planning* 184 (Summer): 13–16.

Williams, Gareth. 2011. *The Mexican Exception: Sovereignty, Police, and Democracy*. New York: Palgrave Macmillan.

Williams, Raymond. 1977. *Marxism and Literature*. Oxford: Oxford University Press.

———. (1958) 2002. "Culture is Ordinary." In *The Everyday Life Reader*, edited by Ben Highmore, 91–100. Reprint: New York: Routledge.

———. (1976) 2015. *Keywords: A Vocabulary of Culture and Society*. Reprint: Oxford: Oxford University Press.

Wilson, David, Jared Wouters, and Dennis Grammenos. 2004. "Successful Protect-Community Discourse: Spatiality and Politics in Chicago's Pilsen Neighborhood." *Environment and Planning A* 36, no. 7: 1173–90.

Wilson, Japhy, and Erik Swyngedouw. 2014."Seeds of Dystopia: Post-Politics and the Return of the Political." In *The Post-Political and Its Discontents: Spaces of Depoliticisation, Spectres of Radical Politics*, edited by Erik Swyngedouw and Japhy Wilson, 1–22. Edinburgh: Edinburgh University Press.

Wright, Melissa. 1999. "The Dialectics of Still Life: Murder, Women, and Maquiladoras." *Public Culture* 11, no. 3: 453–74.

Wyly, Elvin. 2015. "Gentrification on the Planetary Urban Frontier: The Evolution of Turner's Noösphere." *Urban Studies* 52, no. 14: 2515–50.

Young, Iris Marion. 1990. *Justice and the Politics of Difference*. Princeton, NJ: Princeton University Press.

Yusoff, Kathryn. 2018. *A Billion Black Anthropocenes or None*. Minneapolis: University of Minnesota Press.

Zárate, Lorena. 2011. "Mexico City Charter: The Right to Build the City We Dream Of." In *Cities for All: Proposals and Experiences towards the Right to the City*, edited by Ana Sugranyes and Charlotte Mathivet, 263–270. Santiago: Habitat International Coalition.

Ziccardi, Alicia. 1996. "El Estatus de la Capital: Descentralización, Reforma del Estado y Federalismo." *Revista Mexicana de Sociología* 58, no. 3: 99–117.

———. 2014. "Poverty and Urban Inequality: The Case of Mexico City Metropolitan Region." *International Social Science Journal* 65, no. 217–218: 205–19.

———. 2017. "Vivienda, Gobiernos Locales y Gestión Metropolitana." In *Los Gobiernos Locales y las Políticas de Vivienda en México y América Latina*, edited by Alicia Ziccardi and Daniel Cravacuore, 13–30. Buenos Aires: CLACSO.

Žižek, Slavoj. 2005. "Against Human Rights." *New Left Review* 34, (July–August): 115–31.

———. 2008. *Violence: Six Sideways Reflections*. New York: Picador.

———. 2012. *The Pervert's Guide to Ideology*. Film. DVD. Directed by Sophie Fiennes. New York: Zeitgeist Films.

Index

Page numbers in *italic* refer to images.

Instituto Nacional de Estadística y
  Geografía (INEGI), 179, 188n24,
  218n19
Instituto Nacional Electoral (INE), 179
Iveson, Kurt, 101
Iztapalapa, *29*

Jackson, J. B., 9
Jameson, Fredric, 70
jefe de gobierno, 183n1
Johns, Michael, 21
Jones, John Paul, 70
Jornada, La, 40, 93, 123, 178, 215n49
Joseph, Gilbert, 21, 160, 187n15, 193n45
Juárez, Benito, 106, 154, 212–13n22
Juárez Neighborhood School of
  Citizenship, 128–30
Jung, Carl, 70, 207n7

Kanai, J. Miguel, 114
Knight, Alan, 57–58, 156
Kofman, Eleonore, 72
Kristiansen, Annali, 78
Kuri, Rodríguez, 138–39, 185n12,
  198n20

La Trobe, Charles, 185n2
Labastida Ochoa, Francisco, 49, 202n7
Labor Party (Partido de Trabajo, PT),
  205n31
Labyrinth of Solitude (Paz), 194n48
land, redistribution of, 24, 187n19
Landman, Todd, 49
language and meaning, uses of, 7–8,
  184n9
Law of Citizen Participation. *See* Ley
  de Participación Ciudadana
  (LPC)
Leal Martinez, Alejandra, 95
Lear, John, 27, 187n14
Lebas, Elizabeth, 72
Lefebvre, Henri, 9, 68–69, 71–78,
  183n4, 207n16, 207nn11–13
Lenin, Vladimir, 156

leviathan
  Mexico City as, 2
  PRI as, 2, 58, 183n5
  use of term, 183n5
Levy, Simón, 104, 106, 110, 112–20,
  202n21, 213n28
Lewis, Oscar, 190n36
Ley de Participación Ciudadana
  (LPC), 22nn6–7, 89, 98, 103, 112,
  118, 213n30
Linea 12 controversy, 165, 221n1
Lomas, 31, 191n37
Lomnitz, Claudio, 47
López de Santa Anna, Antonio, 21
López Mateos, Ernesto, 187n19
López Obrador, Andrés Manuel
  (AMLO)
  airport construction plans, 165,
    166–67, 189n30
  Cuarta Transformación, 153, 172
  desafuero saga, 203n25
  formation of Morena, 57, 219–20n31
  as GDF mayor, 52–54, 203n23,
    215n49
  presidency, 14, 54–56, 165–66,
    221n1
  and Supervía Poniente, 215n47
  support for, 153–54
López Portillo, José, 37, 38, 40
LPC. *See* Ley de Participación
  Ciudadana (LPC)
Lucumberri Prison, 36–37

Macedo de la Concha, Rafael,
  203–4n25
Madero, Francisco, 22–23, 54, 187n15
Madero, Gustavo, 149
Mancebo, François, 60
Mancera, Miguel Ángel
  and Corredor project, 96, 112–21
  as mayor, 12–13, 56–57, 99, 120, 123,
    208n23, 211n40, 215n48
  and Pacto por México, 133–34, 137–
    38, 162, 217n17

www.ingramcontent.com/pod-product-compliance
Lightning Source LLC
Chambersburg PA
CBHW031415270326
41929CB00010BA/1464